Weather by the Numbers

Transformations: Studies in the History of Science and Technology
Jed Z. Buchwald, general editor

Weather by the Numbers

The Genesis of Modern Meteorology

Kristine C. Harper

The MIT Press
Cambridge, Massachusetts
London, England

© 2008 Massachusetts Institute of Technology

For information on quantity discounts, email special_sales@mitpress.mit.edu.

Set in Stone Sans and Stone Serif by SPi Publisher Services, Puducherry, India. Printed
and bound in the United States of America.

Library of Congress Cataloging-in-Publication Data

Harper, Kristine.
Weather by the numbers : the genesis of modern meteorology / Kristine
C. Harper.
    p.   cm. — (Transformations : studies in the history of science and technology)
Includes bibliographical references and index.
ISBN 978-0-262-08378-2 (hardcover : alk. paper)
1. Meteorology—History. 2. Weather forecasting—History. I. Title.

QC855.H374 2008
551.50973—dc22

2007039860

10 9 8 7 6 5 4 3 2 1

Contents

Acknowledgements

This book is the product of the guidance and assistance of many people, in particular the archivists and librarians who helped me find the materials that I needed to tell this story. With great appreciation I acknowledge the assistance of Diane Rabson, National Center for Atmospheric Research in Boulder, Colorado; Margery Ciarlante, National Archives II in College Park, Maryland; Janice Goldblum, National Academy of Sciences Archives; Judith Goodstein, Charlotte Erwin and Bonnie Ludt, California Institute of Technology Archives; Michelle Blakeslee, Niels Bohr Library, American Institute of Physics, College Park, Maryland; Geoffrey P. Williams, M. E. Grenander Department of Special Collections and Archives, State University of New York, Albany; Nora Murphy, Institute Archives and Special Collections, Massachusetts Institute of Technology; Bradley Gernand and the staff, Manuscript Division, Library of Congress; Deborah Day, Scripps Institution of Oceanography Archives; the archivists of the University of Washington Archives; Jinny Nathans, American Meteorological Society; Pamela Henson, Smithsonian Institution; Doria Grimes, NOAA Central Library; and Joe Toth, Valley Library, Oregon State University. Thanks also to those who helped me find materials and shared information with me while I visited the University of Stockholm: Eva Tiberg, Erland Källén, Karl Grandin, and Anders Carlsson.

I also appreciate the assistance of those who tracked down images for me: Lauren Morone of the National Centers for Environmental Prediction; Air Force historian Gerald White; Liselle Drake, NOAA Central Library; Frank Conahan, MIT Museum; and Alex Nichols and Mandy Altimus Pond, Massillon Museum.

Financial support was provided by the State of Oregon through the Oregon State University External Fellowship Tuition Relief Program and Supplemental Oregon Laurels Graduate Scholarships, the American Meteorological Society's Graduate Fellowship in the History of Science, travel grants from the American Institute of Physics and the National Science Foundation, a postdoctoral fellowship from the Dibner Institute for the History of Science and Technology, and a fellowship from the Tanner Humanities Center, University of Utah.

I especially appreciate the meteorologists who gave their time for interviews: Joanne Simpson, George Haltiner, Fred Decker, Bert Bolin, Willard "Sam" Houston, Leo Clarke, Edward Lorenz, Thomas Malone, and Paul Wolff. The background information I gained from these interviews helped me put the archival information into perspective. Thanks also to Norman Phillips, the late John C. Freeman, and Margaret Smagorinsky—Meteorology Project participants—who offered encouraging words of support, and to Anders Persson of the Swedish Meteorological and Hydrological Institute, who provided information on the Scandinavian part of this story.

Many thanks to my dissertation committee members at Oregon State University who provided comments on early drafts of this manuscript: Ronald E. Doel, Mary Jo Nye, Paul L. Farber, Steven K. Esbensen, and James R. Males.

Encouragement and moral support have come from many people. Particular thanks go to Ron McPherson and William Hooke of the American Meteorological Society. Thanks also to James Fleming, Ronald Rainger, Naomi Oreskes, Roy Goodman, Greg Good, and David Cahan for suggestions and ideas. Many thanks to Gin Gebo and Willa Mae Gird, both of whom provided lodging during archive trips and notes of encouragement. I appreciate very much the support of friends in Albany and Corvallis, Oregon, particularly Carmel Finley, the Rev. Donna Pritchard, the Rev. Barbara Nixon, and members of the First United Methodist Church, Albany, Oregon. Thanks also to my colleagues at New Mexico Tech: Glenda Stewart Langley, who volunteered to be an "interested non-specialist" reader and gamely read and commented on the entire manuscript; and Sue and Doug Dunston, Mary Dezember, Maggie Griffin, Miriam Gustafson, and Ann Hewitt (all of the Humanities Department), and librarian Stephanie Wical—steady boosters all.

This book would not have been possible without the encouragement and support of the wonderful professionals at The MIT Press. My thanks to series editor Jed Buchwald, acquisitions editors Sara Meirowitz and Marguerite Avery, acquisitions assistants Kristan Palmer and Erin Mooney, catalog manager Susan Clark, and digital manuscript and art coordinator Julie Lavoie. Special thanks to senior editor Paul Bethge.

On the home front, I could not have completed this project without the cooperation and full support of my daughter, Teresa Elias, who tended pets and kept the household running while I flew to far-flung archives. Thank you, Teresa, for putting up with the disruption, keeping me supplied with great cookies, and being a terrific wordsmith. Thanks also to my mother, Helen Harper, for her steady encouragement. And to Ron—my partner in all things—who read through this manuscript more times than anyone should have: thanks, Sweetie!

Weather by the Numbers

Introduction

Wednesday, 9 January 1946. As clouds filled the sky and rain began falling through the chilly morning air, a dozen meteorologists gathered in the US Weather Bureau's castle-like headquarters, less than a mile from the White House. By 10:30 a.m., Chief of the Weather Bureau Francis Wilton Reichelderfer, top members of his staff, and several military meteorologists were huddled in a confidential meeting with two eminent guests. One was John von Neumann, a brilliant mathematician from the Institute for Advanced Study in Princeton; the other was Vladimir Zworykin of RCA, inventor of the scanning television camera. The two visitors had come to Washington to discuss their "startling, but *noteworthy* proposal" to use von Neumann's planned electronic digital computer to forecast, and ultimately to *control,* the weather.[1]

Although Reichelderfer, von Neumann, and Zworykin believed that their meeting was confidential, the *New York Times* broke the story of this proposal two days later. At the time, it was overshadowed by the first session of the United Nations' General Assembly and by a strike by New York's Western Union employees, which crippled the city's communications with the rest of the world. But for the meteorologists of America's weather services—indeed for the global community of meteorologists—it marked a crucial transition. Life in a discipline long maligned as a "guessing science" (as Theodore von Kármán of the California Institute of Technology's Guggenheim Aeronautical Laboratory had put it) was about to be transformed in almost unimaginable ways.[2]

Meteorology has undergone significant disciplinary changes in the past 100 years. Early-twentieth-century meteorologists would be amazed by today's practice of their science. Once an art that depended on an individual forecaster's lifetime of local experience, meteorology has become a sophisticated, *theoretical* atmospheric science. Its data—no longer limited to what can be transmitted over telegraph lines—travel over high-speed links from land-based observing stations and myriad remote sensing instruments around the world.

But data availability was not the sole, or even the primary, reason for the twentieth century's meteorological advances. From the beginning of the century to the immediate post-World War II period, weather forecasters had increasingly large amounts of data landing on their desks via clattering teletypes. But the human mind can successfully process only a limited amount of new information in a short period of time. Forecasters discarded extra data despite their potential usefulness to weather prediction. Dynamicists—students of atmospheric motion—could have used the data to solve their newly developed equations had they had months or years available for calculations. As British meteorologist Lewis Fry Richardson had determined during his famously abortive attempt at numerical weather prediction during World War I, "[64,000 human] computers would be needed to race the weather for the whole globe."[3] The inability to quickly solve nonlinear equations (whose unknown variables are raised to a power greater than one) greatly hampered the development of theories. Therefore, many advances in meteorology, particularly in the latter half of the twentieth century, depended on one technological innovation: the electronic digital computer.

This book tells the story of American meteorology's transformation from a discipline more art than science in the early twentieth century to a sophisticated science by 1955. In 1900, the US Weather Bureau and American meteorology were synonymous. "With a few notable exceptions," *Scientific American* noted in 1911, "[the Weather Bureau's] personnel include all of the professional meteorologists in the country."[4] No other American scientific community was so closely tied to a single governmental agency. Federal dominance exacerbated, and was exacerbated by, a chronic lack of funds for applied and theoretical research, a severely mathematics-deficient general circulation theory, a dearth of academically trained practitioners, and an "unscientific" disciplinary reputation among scientists.

The rapid rise of aviation during the Great War triggered a demand for military meteorology—a demand that did not abate during the 1920s and the 1930s.[5] As a direct consequence, academic programs in meteorology arose in the late 1920s and continued to grow, albeit slowly, throughout the 1930s. With Europe experiencing political upheaval in the late 1930s, another major war loomed. Military planners recognized that this war would put even heavier demands on aviation assets. No longer just a means of tracking enemy movements, aircraft would be used to ferry materiel and to deliver weapons. To ensure safety of flight, the military services needed a significantly improved network of surface and upper-air observation stations, and thousands of new meteorologists.[6] An expanded observation network, newly trained mathematics- and physics-savvy meteorologists ready to define the atmosphere in mathematical terms, and the nascent electronic digital computer combined to open the door to a radically new way of approaching both atmospheric theory and weather forecasting: *numerical weather prediction.*

The introduction of numerical weather prediction—the creation of forecast weather maps by computer instead of by people—profoundly changed the science of meteorology. The extension of numerical modeling to other scientific and technological undertakings, including simulated atomic tests and biological population studies, would lead to a fundamental change in conducting scientific research. The Meteorology Project, which developed these early atmospheric models, thus spearheaded the introduction of scientific modeling in the twentieth century.

The Meteorology Project was the sister project to John von Neumann's Computer Project at the Institute for Advanced Study. Previous accounts have made von Neumann and his computer the stars of the Meteorology Project. Frederik Nebeker's *Calculating the Weather* casts numerical weather prediction as one in a long line of calculating techniques aimed at easing the forecasting burden during the twentieth century, while William Aspray's *John von Neumann and the Origins of Modern Computing* presents the Meteorology Project as a small part of the main show: the development of von Neumann's computer.[7]

Both Nebeker's book and Aspray's approach the Meteorology Project and numerical weather prediction from the viewpoint of physics. In a significant and important shift in perspective, my book expands the existing historiography by focusing on meteorologists and the American meteorological community in the twentieth century. It challenges the prevailing interpretation that gives John von Neumann virtually all the credit for the success of numerical weather prediction. That interpretation errs in several respects. Von Neumann's invention of the electronic digital computer was indeed necessary for the rise of numerical weather prediction. But von Neumann alone could not have developed it. Meteorologists formulated the equation sets (models) for von Neumann's computer to solve. In existing accounts, the few meteorologists credited with developing numerical weather prediction were the Americans working in Princeton. The rarely acknowledged "Scandinavian Tag Team" members, rotating in and out of their Norwegian and Swedish posts from 1948 until 1955, played important roles throughout the project's life. However, meteorologists who were occasional visitors to Princeton—most notably Swedish-American Carl-Gustav Rossby, Weather Bureau Chief Francis Reichelderfer and his Scientific Services Division head Harry Wexler, Daniel Rex of the Office of Naval Research, and Philip Thompson of the Air Force Cambridge Geophysical Research Directorate—were the project's real movers and shakers. Therefore, despite being a military-funded Cold War project, this was very much a civilian-influenced international effort conducted at the boundary of theoretical and applied meteorology. With Rossby in the lead, meteorologists knew that their discipline depended on international cooperation no matter what the political situation.

This book also explores the role of national styles in scientific development and practice. This "American story" is full of Scandinavian characters—scientists imported to bridge the gap separating meteorological theorists and operational weather forecasters in the United States. Unlike their American counterparts, Scandinavian meteorologists working for their nations' weather services not only analyzed charts and made forecasts, they also did theoretical research. US Weather Bureau meteorologists, in contrast, had been trained on the job—not in colleges and universities—since the bureau's 1891 establishment under the Department of Agriculture. They made forecasts—they did not do theoretical research. The few academic meteorologists in the United States—and there were very, very few—concentrated on theoretical matters. They did not make forecasts. The differences in national styles became a determining factor in deciding who would staff the Meteorology Project.

Third, this book makes clear the importance of Carl-Gustav Rossby as the leader of a research school that profoundly influenced twentieth-century meteorology. Historical studies of research schools have focused on laboratory sciences. The few that address field sciences do not include atmospheric science. Rossby, though in Stockholm and geographically distant from the project, touched every aspect of it. He influenced from afar, but he was crucial to successful international acceptance of numerical weather prediction techniques as valuable theoretical and forecasting tools.[8]

Last, this book gives a rightful place to the control of nature as a guiding influence for the Meteorology Project. Von Neumann and Zworykin thought that once weather could be predicted it could be easily controlled. They anticipated changing the variables in computer models of the atmosphere until they produced, as one would say today, the desired "virtual" weather. By modifying the same variables in nature, they could move from virtual weather to the real thing. These two non-meteorologists expected to produce weather on demand. As von Neumann argued in the Meteorology Project's Justification Memorandum, "[the Project would take] the first steps toward influencing the weather by rational, human intervention . . . since the effects of any hypothetical intervention will have become calculable."[9] The possibility of controlling the weather—making it a potential offensive and defensive weapon—greatly interested the Meteorology Project's financial patrons: Navy admirals. (Most American meteorologists, however, had no interest in weather control. They were struggling to develop a comprehensive theory and to deliver forecasts to their customers: the American public.) Funding and interest by non-meteorologists in numerical weather prediction stemmed not from the advantages of better and faster forecasts, but from the tantalizing possibilities presented by designer weather. The control of nature was a major theme in the immediate postwar era as scientific, technological, and engineering fixes—for example, vast flood control and irrigation projects—were touted to ameliorate natural conditions that adversely affected people. However, what the supporters of weather control seemed to forget was that it was impossible to control the atmosphere over one patch of ground without affecting the atmosphere over someone else's patch of ground.

The history of numerical weather prediction is, thus, far more than the story of how a small group of meteorologists combined equations of motion, hydrodynamics, and thermodynamics into computer-solvable models of the atmosphere. The development of numerical weather prediction in the United States is a story with multiple themes: the development and professionalization of a scientific discipline, the influence of international science on national practice, the significant role played by a far-flung research school in a non-laboratory science, and the joining together of disparate parts of the same disciplinary community to produce a sharp break with past scientific practice. It is a story of meteorologists and of the role of meteorology in the twentieth-century United States.

Overview

Chapter 1 examines the stagnant state of meteorological services during the interwar period. Although the "meteorological renaissance" of Vilhelm Bjerknes's Bergen School of meteorological theory and practice had taken Europe by storm, the expansion of upper-air observation networks and incorporation of new mathematics-based forecasting techniques did not spread across the Atlantic.[10] The US Army Signal Corps and the US Navy had maintained meteorological services during the Great War, but rapid demobilization left both services with skeleton crews. The Weather Bureau was *the* agency responsible for the nation's weather. Operating on a small budget of $2 million annually—or two cents per person—the bureau's free weather forecasts saved American farmers and businesses tens of millions of dollars by reducing crop losses.[11] The bureau's climatology records, provided at no cost, allowed the insurance industry to expand its offerings of "pluvius weather policies," which guaranteed game receipts to sports promoters in case of bad weather.[12] Despite the Weather Bureau's positive economic influence—not to mention its importance for safety of flight and shipping—its budget and staff shrank due to the funding reductions of the Great Depression. What little progress had been made—primarily in aviation-related support—gave way to drastic retrenchment during the 1930s.

The weather services were not the only sectors of American meteorology in trouble. Just two PhDs were awarded in the discipline between 1919 and 1923—compared to 621 in chemistry, 185 in physics, and 93 in the geological sciences.[13] Because, as Harvard climatologist Robert DeCourcy Ward opined, "everyone is a born meteorologist," few felt inclined to study it.[14] Meteorology languished at the bottom of the academic barrel. Despite the lack of academically trained meteorologists in the United States, this period saw the emergence of academic programs, which had been introduced in Europe in the middle of the nineteenth century. Chapter 2 discusses the establishment of meteorology programs, and later departments, beginning in the late 1920s. Rossby's theoretical program at the Massachusetts Institute of Technology was followed within a few years by applied aviation-related meteorology programs at New York University and at the California Institute of Technology (Caltech). Young people interested in meteorology—not to mention forecasters already employed by the US Weather Bureau, the US Navy, and the US Army Signal Corps—had domestic options for meteorological education. Having formed their own national professional organization in 1919—the American Meteorological Society (AMS)—a higher-profile meteorology community took shape. But this late entry into the realm of professional societies was radically different from other professional groups. In contrast with the American Physical Society, fully half of the AMS's members were *amateurs*. The only requirement for membership was an interest in weather.[15] The AMS encouraged the expansion of educational opportunities at all levels and worked to influence an emerging research agenda in meteorology. The new academic departments along with the AMS nurtured a slow but steady advancement of scientific theory, practice, and education in meteorology during the interwar period. These advances turned out to be valuable for both military needs in World War II and the postwar theory-based efforts to create numerical weather prediction.

The relatively low demand for weather forecasts despite the rapid increase in numbers of aircraft and flight hours left the American meteorological community of approximately 400 professionals extremely unprepared to meet the nation's need for atmospheric support during World War II. Chapter 3 addresses the small academic community's response: the training of about 8,000 meteorologists and 20,000 meteorological observers and technicians under the direction of Rossby's University Meteorological Committee.[16] Composed of one representative from each of the "Big Five" meteorology programs (those at MIT, NYU, Caltech, the University of California at Los Angeles, and the University of Chicago), the committee—with the help of every academic meteorologist in the country—had fulfilled its training mission by 1943 and had begun considering how academic meteorology could influence the postwar research agenda. The primary point of discussion was disciplinary professionalization. With a twentyfold increase in university-trained meteorologists standing ready to influence the community, Rossby and his confreres were determined to seize this rare opportunity to move meteorology from an art to a theoretically based science respected by both the public and their fellow scientists. "It is an unfortunate characteristic of meteorology," University of Chicago meteorologist Horace Byers lamented, "that its great forward strides depend on disasters. [And] World War II was an outstanding example of growth bred from disaster."[17] Channeling that growth in the postwar period would be of great consequence to the rise of numerical weather prediction.

Rossby and his colleagues shared one overriding scientific goal: to pursue basic meteorological research aimed at developing a mathematics-based theory of general circulation. That goal became wedded to John von Neumann's Computer Project in early 1946. How this project came into being is the subject of chapter 4. Francis Reichelderfer was introduced to the idea of computer weather forecasting during a visit to RCA's Princeton labs in 1945. Excited by this possibility, he encouraged von Neumann to use his "electronic brain," then under development, to forecast the weather using numerical methods. Money, as usual, was a major stumbling block. The Weather Bureau controlled almost no research funding—research was not part of its mission. Not easily discouraged, Reichelderfer called upon his long-time colleague Rossby to formulate a plan and arrange a patron for a meteorology project at the Institute for Advanced Study. The entrepreneurial Rossby—who during the war had sent unsolicited telegrams on weather-related matters directly to Vice President Henry A. Wallace, a long-time acquaintance—had many contacts within the highest reaches of government. He quickly secured financial backing from the Office of Naval Research. After arranging funding and personnel, Rossby suggested that von Neumann pursue an approach that would first require theoretical development in meteorology, then follow with an operational application to weather prediction. Von Neumann readily agreed, and by mid 1946 both the Computer Project and the Meteorology Project were underway. But the Meteorology Project was immediately hampered by hiring difficulties that severely limited progress and almost drove the frustrated, despairing von Neumann to abandon weather forecasting.[18] Fortunately, wartime-educated meteorologist Jule Charney, under Rossby's mentorship, was spending most of 1947 studying in Norway and developing a simplified system of equations that would describe atmospheric motion. With his equations and a method to filter out "noise," Charney and the first member of the Scandinavian Tag Team, Arnt Eliassen, were ready to join the Princeton team and turn the "ugly duckling" of meteorology into a "swan."[19] Drawing on extensive archival collections, this chapter reveals the distinctly different character of American and Scandinavian meteorological practice.[20] Knowing that Eliassen's presence on the team would be important for its success, Charney convinced von Neumann that they needed at least some members who had "intimate experience with actual weather processes."[21] The Scandinavians consistently filled those positions. This chapter also provides important new insights into the development and operation of this project, demonstrating that meteorologists, and not von Neumann alone, were the intellectual leaders of this far-reaching enterprise.

When Charney and Eliassen arrived in Princeton, they brought along the methods and influence of Rossby—the de facto head of the Meteorology Project—and his research school. In chapter 5, I discuss how Rossby's research school not only influenced international meteorology generally, but also overcame the initial skepticism of both theoretical and applied meteorologists who doubted that numerical weather prediction was a valid and necessary technique for extending meteorological theory and improving weather forecasting. Founder of two meteorology programs in the United States (those at MIT and the University of Chicago), responsible for wartime training, and founder of the first peer-reviewed meteorological journal in the United States and of a Swedish journal aimed at the broader international geophysics community, Rossby was in the perfect position to convince the international meteorological community of the wisdom of numerical weather prediction. Providing a series of Scandinavian meteorologists who could bridge the gap between synoptic (current weather analysis) and dynamic (atmospheric motions) meteorology, Rossby played a significant and to date largely unheralded role in the successful development of numerical weather prediction techniques.

Four years into the Meteorology Project, Charney and his team were ready to try their simple barotropic model on a computer, but von Neumann's computer remained unfinished. Chapter 6 tells how the Meteorology Project pushed forward anyway, testing a variety of atmospheric models. First run on Army Ordnance's ENIAC and then on von Neumann's computer, the team members modified the models to get a "match" between the computer forecast and the weather that had occurred. The initial post hoc forecasts of the infamous (and embarrassingly unpredicted) 1950 Thanksgiving Day storm, which had dumped large amounts of snow and rain in the Mid-Atlantic states, were made in spring 1952. The results were poor. The first two computer models failed to catch the explosive deepening of the low-pressure system. But the third model—a simple two-level baroclinic model—did the trick. Even though the "predicted" low-pressure center was 240 miles from its observed position, the computer-generated forecast would have allowed forecasters to forecast the rain and snow that had disrupted the Eastern Seaboard.[22] It was a huge "success"—18 months after the fact. The Princeton team members were moving closer toward their goal of creating a realistic atmospheric prediction. They were not the only ones. Army Air Force meteorologist Philip Thompson, an original Meteorology Project member, had been developing and testing his own models at the Air Force's Geophysical Research Directorate in Cambridge, Massachusetts. Alarmed to learn that Charney's group would soon propose real-time operational predictions, Thompson attempted to derail Charney's desire for a joint operational group that would continue to unite the Meteorology Project's weather service participants. Indeed, Thompson wanted to control numerical weather prediction himself under the Air Weather Service umbrella. The resulting dispute embroiled the Weather Bureau, the Navy, the Air Force, and the Meteorology Project. Ultimately, the three weather services continued to guide the project as numerical weather prediction inched slowly toward operational implementation.

Having decided to "go operational," the members of the Meteorology Project team changed their focus (chapter 7) to operational models as they continued their more theoretical work on general atmospheric circulation. The Weather Bureau, pursuing pioneering meteorological methodology for perhaps the first time, concentrated on housing, staffing, and carrying out computer operations as an adjunct to their subjective (hand-drawn) techniques. All three weather services, which were members of the Joint Meteorological Committee of the Joint Chiefs of Staff, coordinated—more or less harmoniously—the details of the new Joint Numerical Weather Prediction Unit (JNWPU) to be located in Suitland, Maryland. Forming an operational organization paid for and staffed by weather services with very different cultures was not easy. Inter-service rivalry aside, setting up a computer-based center in the early 1950s was difficult. Computers were unreliable; in late summer 1952, von Neumann's computer produced only 11 hours of processing time.[23] The Eisenhower administration, seeking to reduce appropriations, was particularly concerned about computer purchases. As the JNWPU's administrative coordinator, the Weather Bureau had to justify its requirement for a single-user computer powerful enough to handle weather forecasting. Additionally, service representatives were forced to execute a competitive bid for a computer in a period of limited competition and answer inquiries from Commerce Department bureaucrats who wanted to know why the Weather Bureau was unable to combine the best attributes of two different computers produced by two different companies to create an even better computer.[24] Despite the challenges (those externally created and those internally induced), the Air Force, the Navy, and the Weather Bureau produced their first "operational" weather map in May 1955—almost 3 years after deciding to move numerical weather prediction from the realm of research to operations. In so doing, they advanced numerical techniques more quickly than would have been possible in the less time-critical research environment.

The opening of the JNWPU marked the end of the preliminary research period, but it was just the beginning of the worldwide spread of numerical weather prediction. Chapter 8 briefly extends the story to the current time. As their very different meteorological missions exacerbated cultural differences, the Navy and the Air Force removed their personnel and formed their own operational prediction units, leaving the Weather Bureau to fund and staff its own center. As computer availability, processing speed, and memory capacity increased, universities began their own modeling and research projects. The modeling and prediction efforts of individual European nations joined forces to create the European Center for Mid-Range Weather Forecasting (ECMWF), which would provide formidable competition to US-based efforts. In time, modelers would attempt to forecast for longer and longer periods until long-range forecasts took the first steps to becoming climate models.[25]

It has been 50 years since the first operational forecasts made their appearance. Just barely satisfactory by the standards of their time, they would be even less satisfactory today. But they were a start. And the modeling of the atmosphere—for both operational purposes and theory development—continues, and will continue, expanding the knowledge of all those who seek to understand the atmosphere's secrets.

1 :: A Stagnant Atmosphere: The Weather Services before World War II

The meteorological "renaissance" that began in Norway and spread to other European countries at the close of World War I did not extend to the United States. In Europe, meteorology held the same "rank" as astronomy in academic institutions, and research on its theoretical underpinnings was carried out at several academic institutions in Norway, Germany, and England. But in the United States, the top academic institutions did not treat meteorology as a topic on a par with any physical science. If offered, meteorology was typically found in geography courses covering climate. At state universities, meteorology courses were often related to agricultural instruction and developed by Weather Bureau personnel assigned to conduct state climatological and crop studies.[1]

The desultory status of meteorology in the United States, where other branches of the earth sciences began growing rapidly in the first part of the twentieth century, was particularly pronounced. Academic geophysics benefited from new philanthropic support. In 1905, the newly founded Carnegie Institution in Washington launched its new Geophysical Laboratory to "take possession of the vacant ground between geology and physics and geology and chemistry," and its $2 million endowment soon made it an international leader in petrological studies.[2] In 1909 the Carnegie Institution's Division of Terrestrial Magnetism, boldly declaring its intention to map the geomagnetic field of the entire earth, commissioned the non-metallic ocean-going ship *Carnegie* to undertake this survey. A subsequent Carnegie Institution grant established the Seismological Laboratory at the nascent California Institute of Technology in 1921, and that same year the eminent physicist Robert Andrews Millikan was named Chairman of the Executive Council of Caltech.[3] By that time, the Rockefeller Foundation's General Education Board also was supporting research in academic geophysics, offering a grant to Harvard University to support experimental physicist Percy Bridgman's studies of high-pressure on materials. The appointment of Norwegian oceanographer Harald Sverdrup as director of the Scripps Institution of Oceanography in 1936 signaled its rise as a leading research center in physical oceanography.[4] Though as late as 1940 no US university offered a curriculum in all the component fields of geophysics recognized by the American Geophysical Union (and critics decried the absence of rigorous mathematics, physics, and chemistry training in most university geology programs), research and PhD production in many fields of geophysics, apart from meteorology, were robust before World War II.[5] Funding remained inadequate for US government agencies involved in geophysics and the earth sciences: for instance, the superintendent of the US Coast and Geodetic Survey complained in 1921 that the salaries for its employees were "below those paid for skilled labor in mechanical trades outside the US government." But the Coast and Geodetic Survey faced fewer challenges in meeting its mission than did government meteorologists.[6] Expanding opportunities for research in other earth sciences fields thus made the contrast with meteorology stronger still.

Meteorological research in the United States was limited because meteorology fell under the purview of the Weather Bureau, which in turn operated under the jurisdiction of the Department of Agriculture. Although the Weather Bureau—headquartered in Washington, DC (figure 1.1)—had a mission to keep the general public informed of upcoming weather conditions, its primary obligation was to provide agricultural forecasts. Because it was a government agency, any research it performed had to produce an immediate practical result.[7] Similarly, the other two very small "weather services" in the country—maintained by the War Department and the Navy Department—existed to provide specialized forecasts for Army and Navy units. Any research they conducted supported operational requirements.

Military use of aviation increased dramatically during the Great War, and with it the importance of meteorology in keeping pilots and aircraft safe. The Weather Bureau received a special appropriation of $100,000 to establish aerological stations and coordinate services with the War Department and the Navy Department once the United States entered the war, and "flying-weather forecasts" for the military and postal service began in December 1918. Although aviation funding continued after the war, the Weather Bureau made little progress in expanding its services during the immediate postwar period. In contrast, European governments were heavily subsidizing the establishment of civil airways and the meteorological services that supported them.[8] As Secretary Charles F. Brooks of the American Meteorological Society (AMS) noted in 1922, the Belgians were "astonished" that the Weather Bureau's annual budget was only $2 million—or two cents per US resident—and he concluded that "meteorological expenditures and general

Figure 1.1
US Weather Bureau Building, Washington, circa 1912. (NOAA National Weather Service Collection, courtesy of NOAA Central Library)

interest in meteorology are greater in Europe than in the United States."[9] Because of the superior financial support, atmospheric studies in Europe were aided by both academic and applied meteorologists.

While meteorology flourished in Europe after World War I, it stagnated in the United States. The initial promise of increased funding, the rise of aeronautics, and the demand for meteorologists that emerged during the war years, very quickly gave way to retrenchment. Progress was limited— academically, theoretically, and within the applied sector. Underfunded, undermanned, undertrained, and chronically discouraged Weather Bureau personnel advanced the practical, forecasting side of meteorology, despite being crippled by externally imposed limitations. With demobilization, the Army Signal Service and the Navy's weather services struggled to provide weather forecasts with wartime leftovers who saw potential career opportunities in flight forecasting for military pilots. And while the Signal Service concentrated on designing and building new meteorological instruments, it was the Navy that actively sought a more theoretical path toward weather forecasting. The Navy's drive to professionalize its ranks would lead to the first graduate meteorology program in the United States. And by the late 1930s, major meteorology programs would be established at MIT, at NYU, and at Caltech. These programs, and others that followed, would lay the groundwork for US meteorology during World War II and the Cold War. This educational foundation was a necessary condition for numerical weather prediction efforts that would begin immediately after World War II.

Weather for All Reasons: The Weather Bureau

The US Weather Bureau became the nation's official weather service in 1891. However, a weather observation network had been in place since the early nineteenth century, when the US Army Medical Department, academics in New England and New York, and the General Land Office began systematically collecting data. By the 1840s, observations had expanded beyond basic temperature, pressure, and precipitation readings to include data on storms and winds. Between 1849 and 1861, the Smithsonian Institution was home to meteorological research, which was directed by the institution's first secretary, Joseph Henry. The Smithsonian Meteorological Project, focusing on storm movement and climate statistics, was undertaken with several federal agencies as well as the Canadian government. In 1870, the US Army Signal Office began telegraphing daily reports of current conditions and forecasts (called "probabilities"), and the Signal Office continued to function as the national meteorological service until 1 July 1891. An act of Congress dated 1 October 1890 (26 Statutes at Large, 653) then transferred weather duties to the Weather Bureau under the Department of Agriculture. The Weather Bureau's functions, as set forth in section 3 of the act, were as follows:

The Chief of the Weather Bureau, under the direction of the Secretary of Agriculture, shall have charge of forecasting the weather; the issue of storm warnings; the display of weather and flood signals for the benefit of agriculture, commerce and navigation; the gauging and reporting of rivers; the maintenance and operation of seacoast telegraph lines and the collection and transmission of marine intelligence for the benefit of commerce and navigation; the reporting of temperature and rainfall conditions for the cotton interests; the display of frost, cold-wave, and other signals; the distribution of meteorological information in the interest of agriculture and commerce and the taking of such meteorological observations as may be necessary to establish and record the climatic conditions of the United States, or are essential for the proper execution of the foregoing duties.[10]

Sixteen divisions within the bureau carried out its mission. Some were administrative (stations and accounts, supplies, printing, telegraph, library); the remainder covered the range of relevant scientific interests—meteorology, hydrology, seismology, volcanology—and the instrument division that supported them.

Across the nation, five regional districts issued forecasts and warnings. The eastern region's office was within the Weather Bureau's headquarters in Washington, and its chief forecaster could veto forecasts issued by the other regions. The eastern region's office also issued the daily weather maps (figure 1.2). The district offices were, in turn, supported by more than 200 regular stations (figure 1.3), each employing between one and fifteen full-time paid employees who took and transmitted observations and issued local area forecasts. If these employees had time, they performed supervisory functions and conducted limited research. Repair stations and vessel reporting stations also had full-time paid staffs. Additionally, nominally paid ($10–$25 per month) employees made specific observations (for example, by reading river gauges). Since these stations could not adequately cover the nation, several thousand unpaid volunteers maintained "cooperative stations" and collected data for climatological studies and for crop and road services. These volunteers often distributed forecasts and warnings in their local area.[11]

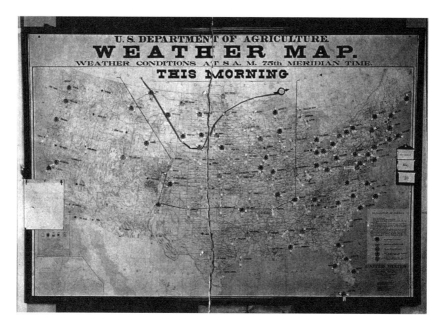

Figure 1.2
Weather Bureau weather map, circa 1900. (NOAA National Weather Service Collection, courtesy of NOAA Central Library)

Figure 1.3
Unknown Weather Bureau office, circa 1900. (NOAA National Weather Service Collection, courtesy of NOAA Central Library)

Paid or volunteer, Weather Bureau personnel were dedicated to providing the best possible weather forecasts to a wide variety of agricultural, commercial, and industrial interests. Although many people thought the recently inaugurated (1919) and highly publicized aviation service occupied the bulk of the bureau's time, in fact it was a minor, albeit growing, portion of the workload.[12] Furthermore, since many citizens were involved, directly or indirectly, with agriculture, it is easy to understand why many people thought the Weather Bureau provided services just to them.

By the early 1920s, the Weather Bureau's five regional offices produced weather maps and written forecasts for the general public, and transmitted them to major media outlets. Newspapers in larger communities printed the forecasts that were also posted in a variety of public places: railroad stations, post offices, hotels, and department stores. The bureau was *the* source for weather information. Local stations issued forecasts and severe weather warnings for a 20-mile radius.[13]

The Weather Bureau also performed extensive work in agricultural meteorology. Although most forecasts and advisories were tailored to a single crop, the weekly *National Weather and Crop Bulletin* presented the previous week's meteorological data and the weather's effect on vegetation, stock, and farm work. The bureau collected specialized data for corn, wheat, cotton, sugar, and rice states. (Cattle-grazing states pushed hard for similarly tailored information.) It published data on fruit frost for tobacco, fruit, truck, and alfalfa seed districts. Fruit-frost warnings were important to citrus growers in California and to orchardists in Oregon and Washington—such warnings enabled them to heat their orchards by burning oil in smudge pots and thereby save their crops from a hard freeze. Similar forecasts and advisories were aimed at tobacco and grain farmers, at New York apple growers (who needed to spray for scab), at millers (who needed to rid their mills of Mediterranean flour moths by flooding them with very cold air), and at beekeepers (whose bees needed a cleansing flight). Western foresters lobbied hard for expansion of fire-weather forecasting, which used meteorological conditions to warn of extreme fire danger and to advise when precipitation would help to quench fires. Additional appropriated funds, in combination with private funding, helped to develop and extend more detailed warnings in fire-sensitive areas.[14]

The Weather Bureau's routine use of probabilities for agricultural and more general forecasts for the public led to a widely held assumption that meteorology was more statistical than geophysical. For example, environmental historian Stephen J. Pyne later argued that meteorology is "a statistical science" because it deals with large-scale events. He further declared that fire had "helped to bring meteorology out of the clouds and back to the earth."[15] However, both assessments are incorrect. Although statistical methods were used to draw information from long-term climatological trends and to develop forecast probabilities (e.g., a 50 percent chance of rain), meteorology was and is fundamentally a geophysical science. Furthermore, the field forecasters—the ones providing fire-weather forecasts—have always been "down to earth." Forecast centers, at least in the United States during the first half of the twentieth century, had no time for theoretical flights of fancy.

Supporting commerce, Weather Bureau forecasters advised shippers when extreme temperatures might harm produce and animals in shipment. For example, freezing temperatures could ruin bananas in transit, and extreme heat could kill livestock en route from farms and ranches to feedlots and slaughterhouses. As Chicago Weather Bureau "official forecaster" Henry J. Cox wrote for *The American Magazine*, "the weather has a finger, so to speak, in almost every business pie." Cox emphasized that businesses dealing in perishable crops and livestock would do well to consult the weather map or call their local forecast office for advice. Doing so saved businesses millions of dollars every year. If it were not for the Weather Bureau, Cox continued, consumers "would have to pay more for [their] fruit and vegetables."[16] These free weather forecasts saved businessmen and their customers millions of dollars annually—many times the bureau's budget.[17]

The Weather Bureau also created "highway forecasts," which had been demanded by automobile associations and by road commissioners who needed to know when to activate snowplows during the winter. Marine forecasting also fell to the Weather Bureau, which in 1904 assumed this responsibility from the Navy. Cooperative agreements with ocean shippers, including the fleets of Standard Oil and the Texas Company, enabled the bureau to receive timely observations from ocean areas, which assisted forecasters making predictions for these same units.[18]

Not all of the Weather Bureau's tasks were meteorological. Utilities and businesses that used fresh water encouraged the bureau to expand river and flood services. Flood warning systems were adequate, but precipitation-related streamflow measurements remained underfunded—a significant deficiency during the great drought of the early 1930s. Additionally, the bureau began collecting and publishing earthquake data in 1914 and monitored volcanic activity—especially on Hawaii.[19] Although the latter task was eventually passed to the US Geological Survey, the Weather Bureau was apparently viewed as an all-purpose collector of earth science data, weather-related or not.

The fastest-growing forecasting and data-collection effort during the interwar period supported aviation. Aeronautics—airships, balloons, and fixed-wing aircraft—had taken on greater importance because of World War I. At war's end, military meteorological organizations that had expanded to fill the need rapidly contracted. However, the aviation assets remained, and safe flight required accurate forecasts of take-off, in-flight, and landing conditions. The undermanned military services could not provide the forecasts, and increasing numbers of air mail flights also required weather information, so in July 1919 the bureau started its flying weather forecasting service for the Army Air Service, the Navy, and the Postal Service. Shortly thereafter, commercial aviation companies began requesting forecasts. Aviators also made more frequent pre-flight visits to weather stations.

That pilots wanted forecasts was good news. But forecasters did not have sufficient upper-air observations and local surface observations to provide them with useful weather information. Demand increased annually as larger numbers of pilots requested forecasts and other specific weather information. The Weather Bureau negotiated cooperative agreements with both the Army Air Service and the Navy. Air Service pilots began visiting stations to obtain weather information and to make contacts with supporting meteorologists, and the Weather Bureau initiated a lecture program that touched on climatology, air currents, physics of the air, and other aviation-related meteorological topics. Additionally, a Navy meteorologist had a desk at the Weather Bureau's headquarters office so he could prepare a weather map with his counterparts there before transmitting it to naval installations throughout the nation. The Army and the Navy shared the cost of obtaining upper-air data by forwarding their pilot-balloon (PIBAL) reports to Weather Bureau headquarters. The (usually) red basketball-sized pilot balloons, filled with hydrogen gas, were sent aloft and rose at a predetermined rate. (At night, observers hung a small paper lantern containing a lighted candle well below the balloon so they could track it in the dark.) Observers tracking the balloon could then determine winds aloft near the station. Kites also carried meteorological instruments that recorded upper-air information (figure 1.4).[20]

The rapid growth in aviation after World War I had a huge impact on the Weather Bureau. By the early 1930s, the United States had 25,000 miles of civil airways for which the bureau provided support, relying on more than 500 cooperative (volunteer) and second-order (minimal pay) stations along the routes. The 13,000 miles of airways that supported all-day flying were served by 24-hour stations. These were significantly more expensive to operate than stations providing routine weather services to agriculture and the public.[21]

Responding to yet other constituencies, the Weather Bureau also began providing climatological data. More than 4,500 volunteers ("cooperative observers") made observations and mailed them monthly to headquarters. The climatology section compiled the data, determined average temperatures and precipitation, and published the results.[22]

During the interwar period, the US insurance industry expanded its product line from life and fire coverage into weather insurance and became a major consumer of climatological data. As this insurance sector grew, so too did the demands on the Weather Bureau for another "free" product. Anyone could ask for—and receive at no charge—climatology data destined for publication. Rainfall data that would not otherwise have been computed could be ordered at a cost of 70 cents an hour in overtime pay.[23] Demand for rain and hail insurance increased dramatically in the 1910s. In New York, Henry W. Ives and Company issued "Pluvius Weather Policies" that insured against losses due to unfavorable weather: crop destruction, penalties due to construction delays, and gate receipts from washed-out sporting events. Customers had to purchase the policy at least a week in advance of the insured event. Why a week? Because weather forecasts were good only for about 24–36 hours. No one could predict the weather a week in advance. Companies based premiums on climatological data. Premiums were higher, for example, if rain

Figure 1.4
"Kite equipped for meteorological observations," circa 1912. (NOAA National Weather Service Collection, courtesy of NOAA Central Library)

was more likely than dry weather on the day of a sporting event. In 1919, customers spent $30 million on premiums covering $500 million in risk.[24] Yet for the climatology data to determine this risk, the insurance companies paid next to nothing.

The Weather Bureau had a clearly defined civil role, but during wartime it also supported military operations. During World War I it provided weather forecasts and observations (surface and upper air) in support of aviators, balloonists, and artillery units. Two members of the Weather Bureau staff took on reserve status in the Army Signal Service and worked with others from the bureau who, along with their British and French counterparts, had joined the active-duty forces to form a special forecasting unit in Europe. In addition to aviation, as detailed below, their forecasting duties included predicting winds for gas dispersal and ballistics.[25]

On the home front, the Weather Bureau provided forecasts and warnings to Army camps and Navy bases, and forecasts to railroads handling food and supplies. The bureau also provided meteorological instruments to the military services and climatological data to the Surgeon General's office, studied upper-air conditions for aviation, reported vessels entering and leaving ports where they had stations, and assisted in organizing gas and flame regiments.[26]

::

After the armistice of November 1918, the question arose of which agency would continue to provide weather services to military activities. President Woodrow Wilson convened a special board that heard arguments in support of and in opposition to separate military meteorology services. The Weather Bureau's leaders vehemently opposed any suggestion that it should not provide all of the nation's weather services. While acknowledging the necessity of maintaining a small number of trained personnel serving meteorological units at flying fields, naval bases, and ordnance proving grounds, the bureau argued that the United States had too few qualified meteorologists to spread them among several agencies. With 90 percent of the "trained and dependable" meteorologists in or associated with the Weather Bureau, its leaders argued that in the event of another war the US government should do what the Coast Guard had done during World War I: make Weather Bureau personnel members of the military services.[27] The provision of weather services was an ongoing point of discussion that extended through World War II and into the Cold War era as the government sought to eliminate duplication of services.

The impoverished Weather Bureau, already stretched thin just trying to meet the myriad demands of its non-paying customers, was not in a position to pursue a research agenda. In contrast, many Department of Agriculture agencies devoted considerable time, talent, and funding to research. While 18 percent of the budget of the Bureau of Chemistry and Soils and almost half of the budget of the Bureau of Experiment Stations were earmarked for research, the Weather Bureau's budget included *no* research funds.[28] Weather Bureau appropriations covered routine weather services for the public, for agriculture, and for industries. Research efforts came last, if at all.

Weather Bureau meteorologists were interested in advancing their discipline, but their "investigations" did not usually extend to asking or answering theoretical questions. Investigations focused instead on the relationship between weather and crops, on storm development, on upper-air conditions, on climatology, on how solar radiation affected weather and climate, and on the improvement of meteorological instruments. With the government's emphasis on practical value, Congress was not going to appropriate funds for research not expected to yield economically important results.[29]

Research in agricultural meteorology did include researching the effects of temperature, precipitation, and other elements of weather. Winter wheat is affected by ambient air temperature and by whether precipitation falls as snow or as rain. Determining optimum weather conditions helped farmers anticipate bumper crops or poor harvests. The Weather Bureau continued conducting research on the impact of certain weather conditions on harvests and on the geographical distribution of farm products.[30]

Another Weather Bureau study, one that was important to civil engineers, examined sky brightness—that is, the amount of expected natural lighting during the seasons, at various hours of the day, and under various atmospheric conditions. When designing and constructing buildings, engineers had to make allowance for natural illumination. Although it was beyond the scope of their studies, bureau meteorologists recognized a need to determine and add the amount of light being reflected from surrounding buildings.[31]

Investigations of solar radiation, which became routine in the early 1920s and extended throughout the interwar period, soon embroiled the Weather Bureau in a very public controversy with non-meteorologists. (By "non-meteorologists" I mean scientists who were not engaged in studies of atmospheric physics, not physicists and other researchers who were devoting attention to the atmosphere and to the broader field of geophysics.) This would not be the first or the only time that scientists without a meteorological background would attempt to tell the meteorology community in general and the Weather Bureau in particular what physical variables were really important in understanding the atmosphere.

Scientists both within and outside the Weather Bureau were attempting to determine whether there was a link between solar intensity and weather phenomena that would aid forecasting. Investigating solar radiation involved making continuous records of the amount of radiation received on a horizontal surface to determine the heating rate during the day, the amount of heat lost during the night, and the relationship between these values and atmospheric conditions. Early on, Weather Bureau researchers were not as optimistic as those who argued that observed solar intensity was an accurate indicator of incoming weather, although they did allow that there might be a connection between solar intensity and variations of the weather over many years. Although the Weather Bureau acknowledged that solar radiation was important to weather, it did not believe that insolation (incoming solar radiation) varied greatly from day to day. The variations were so small that measurable meteorological effects were in doubt.[32]

Vigorously and publicly opposing this view was Charles Greely Abbot, Assistant Secretary of the Smithsonian Institution. Abbot, trained in chemistry and physics at MIT, was the second director of the Smithsonian's Astrophysical Observatory. He was convinced that even small changes in the sun's heat could significantly affect earth's weather. By correlating solar output with weather conditions over a period of several years, Abbot became convinced that it would be possible to use measurements of solar radiation alone to predict the weather—and not just for the next day or so, but for weeks, months, or years in advance. Obtaining accurate measurements was the primary difficulty because, according to Abbot, a change of 1 percent in solar radiation received at the earth's surface could produce noticeable weather effects. Therefore, measurement stations were moved from the United States to an observatory in Chile's Nitrate Desert, a place of clear skies and little rainfall. (Another station was established later on Mount Harqua Hala in arid Arizona.) Abbot argued that, whereas the recorded change in solar radiation and weather might be small at the Chilean station because the affected ground area was so large, the same radiation could produce huge changes toward the poles. Therefore, it was not necessary to measure the insolation at the site of the forecasted change—one only needed to get an accurate measurement at a few optimally placed stations.[33]

Senior personnel at the Weather Bureau, particularly Chief Charles F. Marvin, disagreed vehemently. Disputing Abbot's claims, Marvin (an instrument specialist) argued that Abbot's observed "variations" in solar radiation measurements were not necessarily due to changes in the sun's output. Since the measurements were taken after the sun's rays had passed through 20 miles of earth's atmosphere, it seemed more likely to him that radiation variability depended on the state of the atmosphere and not on solar variability. Abbot countered that it made no difference whether the changes were on "the sun, the earth, or some distant star" if they enabled weather prediction.[34] Although a reluctant Marvin agreed to collaborate with Abbot within the limits of the bureau's resources, he clearly thought the entire idea of forecasting the weather on the basis of solar measurements in a South American desert was absurd.

By fall 1926 this controversy had grown hotter. Abbot claimed that it was possible to predict the weather a year in advance by his solar radiation method. Articles in the popular press implied that the "fundamentalists" running the Weather Bureau were just too conservative to embrace this revolutionary forecasting method. As John Billings Jr. wrote for *The Independent*, "[Abbot's] pioneering with solar radiation forecasts has set the tom-toms of the conservative meteorologists beating wildly. The official Weather Bureau, plodding along carefully with day-to-day forecasts . . . would quickly crush [this theory] out of existence." Marvin became so agitated while dealing with journalists over this controversy that his boss, Secretary of Agriculture William Marion Jardine, ordered him to stop talking to the press and "[observe] the dignified silence compatible with [your] official position."[35] Despite Abbot's arguments, Marvin refused to introduce solar radiation measurement as a forecasting technique until scientific evidence directly linked solar radiation changes and identifiable weather patterns. In the meantime, the idea that the Weather Bureau was a reactionary organization became more deeply entrenched in the American public's psyche. Even though the bureau eventually showed that Abbot's correlations had been due to seasonal variations in stratospheric ozone concentration, Abbot remained immune to the bureau's criticism.[36]

Abbot was not alone in promoting the influence of heavenly bodies on the weather. To the consternation of Weather Bureau leaders, some advocates of that notion were hired by the Department of Agriculture. In October 1934, Secretary of Agriculture Henry Wallace hired Larry Page, a statistician from Wallace's home state of Iowa, to conduct studies of the moon and the stars. Page, who considered stars the "key" to the weather, was appointed to find how these extraterrestrial bodies could aid long-range weather forecasting.[37] Spending money that the Weather Bureau did not have on an idea that meteorologists considered cockamamie must have demoralized the entire forecasting section, not to mention the Weather Bureau's new chief, Willis Ray Gregg, who took over from Marvin in 1934.

Despite arguments over using solar radiation measurements for forecasting, there was no disagreement over their use for agricultural purposes. The bureau investigated the effect of shade cloth used by farmers, i.e., how much radiation needed by plants penetrated different cloth types, and the amount of heat generated by orchard heaters to prevent citrus and other orchard crops from freezing. The bureau conducted experiments on orchard heating (smudging) with the Army Chemical Warfare Service to determine if the smoke barrages, which the Army used to cover troop movements in the field, were effective against frost damage. The studies showed that the best way to protect vegetation from frost was to heat the surface layer by burning the cheapest fuel available.[38] These two investigations were directly related to preserving the economic value of agricultural commodities.

As noted above, the bureau was also responsible for monitoring volcanoes and earthquakes; seismological studies were mandated because they bore a "sufficiently close relation" to what the bureau did with the weather studies. Assigned to take on earthquake duties in 1914, the bureau's mission with respect to earthquake studies and observations was to find and map fault lines and reduce damage to dams, bridges, and other structures by recommending locations away from potential slippage areas. Volcano studies included measuring lava flows and examining the compositions and reactions of volcanic gases. The bureau sought to determine any relation that might exist between volcanic activity and earthquakes, and between volcanic emissions and air and water. It also investigated if volcanic energy could be made available "for the use of man." Relief came in 1924 when the US Geological Survey took responsibility for volcano studies.[39]

Most, but not all, of the bureau's climatological research was related to agriculture. One specific climatological study undertaken at the request of "other departments of the National government and for the use of the [1919] Peace Conference in Paris" dealt with Africa's climate. In particular, the Weather Bureau was assigned to prepare a summary of African climate with special attention paid to former German colonies. The summaries included graphs of monthly precipitation and temperature values for the entire continent as well as a discussion of general climatic characteristics. What department made this intriguing request, or why, was not recorded, but the request illuminated the wide range of demands on bureau resources.[40]

Of all its "research" tasks, however, the primary one was always forecasting—the improvement of short-term forecasts and the extension of the forecast period. The Weather Bureau routinely received requests "from all sides" for forecasts extending months, seasons, and years ahead.[41] A Weather Bureau forecaster assigned to the station at Kansas City, Missouri, reported that he was once asked—*in the winter*—to name a date six weeks in advance when the "sun would shine and [the weather would] be otherwise pleasant" for a bridge dedication. He did so based on climatological information and, by the kind of miracle occasionally bestowed upon weather forecasters, got it right.[42] However, the forecast had "no skill": it was no better than the state of climatology at that time. Skillful, accurate long-term forecasts would be of huge benefit to many business sectors. Farmers wanted advance knowledge of drought, excess precipitation, and extremely high or low temperatures. Road crews and transportation industries wanted to anticipate especially bad winters that could affect their ability to keep goods moving. Manufacturers wanted lead time to produce items needed by consumers. Retail outlets wanted to know what they should order.

However, the bureau's leaders were steadfast in noting that there were, to their knowledge, no "sound physical laws" which would allow such forecasts to be made with any degree of success. This was made more complicated by those outside the science of meteorology and related fields such as atmospheric physics, sociology, and geology who claimed to have discovered methods of making accurate long-term forecasts. Even an economist fancied himself a long-range weather forecaster. The father of econometrics, Henry Ludwell Moore, published a long article in the *Quarterly Journal of Economics* that argued that the eight-year generating cycle in England, the eight-year crop cycles in England, France, and the United States, and the eight-year meteorological cycles could all be tied back to the motion of Venus with respect to the earth and the sun. The bureau was left in the position of sorting through public demands for long-range forecasts based on questionable methods that, upon closer inspection, did not yield valid forecasts. While not denying that it was possible to eventually make such forecasts, it did argue that forecast periods would not increase without solid scientific research.[43]

::

Well into the 1930s, the Weather Bureau doggedly defended its stance against long-range forecasts made without scientific underpinnings acceptable to the meteorology community—i.e., forecasting methodologies that did not include the physical processes of the atmosphere. Chief Charles Marvin's 1930–31 report categorically stated that there was no "real way" to make long-range forecasts. The bureau was familiar with the literature on the subject. Available methods could be categorized as (1) examinations of physical processes that would lead to a specific weather condition, (2) periodicities or cyclical recurrences that correlated astronomical or other sequences of events with a specific weather event, or (3) mathematical correlations between current weather in one location and weather that had occurred in the past, in the same or different location. None of these methods had resulted in any "skilled" techniques that extended the forecast period. A "skilled" forecast, by definition, had to be better than one derived from climatology data and persistence, i.e., the current day's weather would persist into the next.[44]

Bureau officials admitted that they had made very little progress in forecasting the weather in many years. By using radio to improve observational data transmission and airplane observations to gain knowledge of the atmosphere's vertical structure, bureau personnel hoped to expand their understanding of atmospheric dynamics that would aid in attacking the forecasting problem.[45] But unbeknownst to bureau officials, a political storm was brewing on the horizon that would profoundly affect their operation.

The Weather Bureau's leaders knew there were functional areas needing improvement, but viewed their work as being the best their budget allowed. The American Society of Civil Engineers, however, was not content with the services received by the engineering community. In April 1931, the ASCE's Board of Directors appointed a special committee to "give thought as to how the United States Weather Bureau could be made of greater service to engineers." The five-member committee presented its report at the ASCE Annual Meeting held 18 January 1933, and published the report in the January 1933 issue of the *Proceedings of the American Society of Civil Engineers*. A hot blast, their extremely detailed report laid out deficiencies the engineers had seen in the bureau's operation of meteorological observation stations. It was followed by a series of letters, pro and con, which appeared in five subsequent *Proceedings* volumes.

The civil engineering report hit a raw nerve, and the ensuing uproar did not die quickly. Its engineers attacked observation station placement, data handling, and the format in which data were made available to engineers. They also impugned the scientific standing of the Weather Bureau, which, they charged, had "not kept pace . . . with research in other lines of science, either pure or applied." After producing a list of recommendations, the committee members issued a final blast, recommending that upon the retirement of Charles Marvin the president of the United States should appoint a new chief from the ranks of those who were experienced administrators and who possessed "broad fundamental science training" and the "rare qualities of mature judgment and progressiveness." Further, the new chief should be a "courageous [and] diplomatic leader, who will release the latent abilities now bound by archaic tradition." There was one additional recommendation: the new chief need not be a meteorologist.[46]

The ASCE was not the only group complaining. The Navy had been stung by the crash of the rigid airship USS *Akron* (ZRS-4) on 4 April 1933. The *Akron* had been operating off the coast of New England when high winds forced it into the water, where it sank. The accident killed 73 men, including the chief of the Bureau of Aeronautics, Rear Admiral William A. Moffett. A joint congressional committee investigated the crash, and the Navy held a court of inquiry to determine the disaster's causes. Since high winds had forced the *Akron* to crash-land in the water, all eyes turned to the data provided by the Weather Bureau. While Navy aerologists (the term used for weather officers) were required to provide aviators with detailed forecasts for periods longer than a day, the bureau's rather vague forecasts were for only 12 hours. More importantly, for several years the Navy Bureau of Aeronautics had been emphasizing the importance of taking four weather observations per day (instead of the Weather Bureau's two) to the Secretary of Agriculture, who had done nothing to increase the number of observations. After the *Akron* disaster, the Navy wanted action, and it was backed by the congressional investigating committee.[47] The Weather Bureau and the Secretary of Agriculture were under extreme pressure to quickly change their operations.

The loss of *Akron* and the ASCE report caused a firestorm that came to envelop not only the Weather Bureau but also Secretary of Agriculture Henry A. Wallace. Wallace (whose father, Henry Cantwell Wallace, had been Secretary of Agriculture in the early 1920s) was a graduate of Iowa State College. He had worked on the family's paper, *Wallace's Farmer*, becoming editor when his father took the Agriculture Department post. Wallace was also a plant geneticist who worked on corn hybridization. He was very interested in the connection between weather and crops, and he had close ties to the Weather Bureau. However, the ASCE report had become a political hot potato. The *Akron* disaster had the president's attention. Wallace had to address the complaints, or he would find himself under fire for supporting a purportedly non-scientific scientific bureau that could not provide the minimal weather support required for aviation safety. Looking for a way out of this potential quagmire, Wallace found a solution: the Science Advisory Board.[48]

Established to study the functions, relationships, and programs of the government's scientific agencies, the Science Advisory Board had been created by President Franklin D. Roosevelt through Executive Order 6238 on 31 July 1933. The nine-member board, operating under the auspices of the National Research Council, came to be chaired by MIT's president, Karl T. Compton. Board members would offer recommendations to increase government agencies' efficiencies, and aid the nation in exploiting its scientific expertise. The board was concerned with this question: "How far should Government itself go in conducting or supporting research or guiding the applications of scientific discoveries?"[49]

Wallace contacted the head of the National Research Council, Isaiah Bowman, asking for help. Bowman recommended that Compton and Robert Millikan (then president of Caltech) serve on an advisory committee dedicated to addressing the Weather Bureau's problems. Bowman (a geographer by training) also suggested that Wallace consider the statistical records kept by the bureau and how they might come to bear on questions about the atmosphere. Wallace, then on the job less than 6 months, was frustrated with the lack of research funding available for the Weather Bureau. He felt "helpless" to answer the criticisms being heaped upon the bureau and, consequently, on the Department of Agriculture. In the darkest days of the Great Depression, and with the nation's farmers needing extensive assistance, Wallace did not have time to be encumbered by the Weather Bureau's flaws. He was therefore enthusiastic about Bowman's idea to bring in "outside meteorological interests" to improve weather services, to advance science, and to bolster the nation's defense.[50]

In late August 1933, at Wallace's request, the Science Advisory Board created a Committee on the Weather Bureau. Because the members of this committee—Millikan (the chairman), Compton, and Bowman—were not meteorologists, Compton asked Charles D. Reed, a meteorologist in the Weather Bureau's office in Des Moines, to consult.[51] Thus, the committee assigned to "assist" the Weather Bureau was unlike any of the others formed to study the government's scientific agencies: it was composed of scientists who were *not* experts in the agency's dominant discipline. Just as the astrophysicists felt entitled to claim that the sun alone determined the weather, two physicists and a geographer believed that they had a better grasp of meteorological practices than did the meteorologists.

The Committee on the Weather Bureau met with Charles Marvin on 26 August 1933. Bowman noted that Marvin—apparently oblivious to the fact that his days as chief were numbered—was "immensely pleased" with the committee's composition and with its mission. Marvin promised Millikan's committee his full cooperation.[52]

The committee quickly homed in on the subject of meteorological research. Beno Gutenberg (a seismologist who had introduced meteorology courses at Caltech under the umbrella of geophysics) and Lieutenant Commander Francis Reichelderfer (the Navy's senior aerologist) had provided the committee written statements emphasizing the importance of introducing air mass analysis methods. This method, introduced by Vilhelm Bjerknes and his son Jacob at the Geophysical Institute in Bergen, Norway (known in scientific circles as the "Bergen School"), had been available since the early 1920s. Reichelderfer had already introduced these techniques to Navy aerologists. While it did not appear that air mass analysis would significantly lengthen the forecast period, all of the committee members nevertheless believed it would lead to increased accuracy.[53]

Wallace anticipated the committee's first report on 1 November. Millikan volunteered to write the first draft. Committee members agreed that, to underscore their concerns about Weather Bureau's structure and about expansion of research opportunities, they needed to make a case for economic benefits that would be favorably received by agriculture, commerce, and aviation. The report would include their recommendations on the adoption of air mass analysis techniques and the full range of Weather Bureau functions. However, committee members did not concur in the report of the American Society of Civil Engineers, which they found lacking (they thought the ASCE had failed to appreciate the Weather Bureau's many responsibilities and the way it carried out its functions, particularly since the engineers' primary complaint was that they needed to reformat the bureau's data to make it useful for *their* purposes). The committee thereby eliminated one of Wallace's concerns: he no longer had to worry about the engineers' narrowly defined complaints.[54]

The committee still had to address the matter of replacing the Weather Bureau's chief. In early October, retired Weather Bureau meteorologist Oliver L. Fassig visited Isaiah Bowman. While Fassig ostensibly wanted to discuss tradewinds (he was working on a study of tradewind flow in Puerto Rico), his real mission was to discuss Charles Marvin's replacement. Fassig argued that no one in the Weather Bureau had ever encouraged research. To his way of thinking, the bureau still suffered from "the old army spirit" from which it had sprung. The bureau needed someone from outside to come in. Fassig, however, could only think of one person he would recommend to be the new chief: Willis R. Gregg, a longtime Weather Bureau meteorologist. Perhaps more importantly, Fassig was worried that political influences could lead to a choice that ultimately would be detrimental to the bureau's best interests.[55]

By mid November, the committee had a preliminary report (which did not include a recommended replacement for Marvin) ready for Secretary Wallace. The report's primary recommendation was that the Weather Bureau adopt the Bergen School's methods of air mass analysis immediately, with Army and Navy assistance. The Weather Bureau needed the military services' cooperation to expand the upper-air observation system (which had widely scattered stations; see figure 1.5) within its limited appropriated funds. The report also recommended that all data reporting and recording be assigned to the Weather Bureau. To fulfill these recommendations, the bureau needed

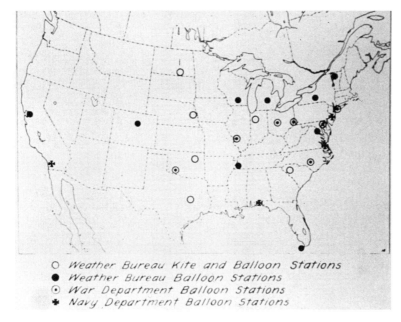

○ *Weather Bureau Kite and Balloon Stations*
● *Weather Bureau Balloon Stations*
⊙ *War Department Balloon Stations*
✳ *Navy Department Balloon Stations*

Figure 1.5
"Kite and balloon stations in the US," circa 1925. (NOAA National Weather Service Collection, courtesy of NOAA Central Library)

to find and hire meteorologists who had training and experience in air mass analysis. It also needed nationwide daily reports of temperature and humidity up to 4 miles above the earth's surface, as well as more frequent and detailed surface reports from both terrestrial and oceanic stations. However, improving weather services was not just a matter of more people and more observations. Even if the people and the reports were in place immediately (an impossibility due to Depression-driven across-the-board funding cuts), the bureau estimated that it would take 3–5 years to introduce the techniques to experienced forecasters. Army and Navy stations would provide additional upper-air data. Obtaining additional surface reports would be difficult: stations were manned to report only twice daily, but at least four daily reports, taken simultaneously around the country, would be needed to produce the four daily maps dictated by the Norwegian method. Congress had not appropriated additional money for expanded data collection, so the bureau could only hope to make limited progress with the new forecasting techniques. If it did not study and vigorously apply the results of new scientific work, the bureau realized, it would fall hopelessly behind other similar institutions. Indeed, it already had. European governments were expending more money than the United States on research and applications, and the Bergen School's techniques were already successfully in use on the Continent.[56]

While not heavily engaged in what would normally be called research, the Weather Bureau was responsible for publishing the only scientific journal in the United States devoted to meteorological research: *Monthly Weather Review*. In addition to publishing articles on scientific advances at home and abroad, *MWR* published recent and average weather records. The Weather Bureau also aimed to eradicate widely held "false ideas, which everywhere abound respecting the weather" and to assist those providing meteorological instruction in secondary schools and in higher education institutions.[57] Additionally, *MWR* fulfilled America's obligation to the wider international meteorological community by providing observational and statistical data related to meteorology and climatology. In return, the bureau received similar information from other nations. *MWR* published investigations of upper-air phenomena (including the strength and direction of air currents), articles on protecting agricultural products from weather extremes, and articles on the role of weather in health-related matters (termed "physiological meteorology"). *MWR* was the only journal publishing fairly long articles on meteorological research. Such studies vanished when *MWR* was crippled by Depression-era funding reductions. In 1932, *MWR*'s editor, the atmospheric physicist William J. Humphreys, stopped publishing articles as a cost-cutting measure—only the data portions remained in the journal. This action temporarily eliminated the one medium for exchanging new meteorological information worldwide, further hindering disciplinary advancement. Funds were restored almost a year later, at which point Humphreys requested immediate submission of completed articles.[58]

Monthly Weather Review was only one of many research-related line items that were cut. In 1932, the entire government research budget was reduced by 12.5 percent. That included the Weather Bureau's scientific work not falling directly under the heading of research.[59] This loss of funding directly affected the bureau's ability to pursue climatological work.

Even more problematic than its paltry research budget was the Weather Bureau's inability to hire and keep scientifically trained staff members. Although the bureau's personnel situation most certainly deteriorated during the Depression, it had been plagued by personnel shortages for many years. Indeed, the War Department and the Navy Department had had few personnel trained in meteorology before the United States entered World War I. Since the majority of people with meteorological training (professional and technical) worked for the Weather Bureau, it had been responsible for providing both personnel and training to the war effort. Despite the resulting increase in demand for services, starting in 1914 the bureau had experienced a *decline* in personnel, even as foreign meteorological bureaus were growing. Funding had not kept up with expenses or the expansion of services. Congress had turned down a request for additional fiscal year 1921 appropriations to cover aerological work in support of military and civil aeronautics, data gathering, and forecasting in support of marine meteorology (the bureau was responsible for open-ocean forecasting), and data gathering and forecasting related to fire-weather, fruit-frost, and other specialized agricultural-related missions. As the Weather Bureau's chief declared in his 1919–20 annual report, stagnant appropriations coupled with rapidly rising costs for goods and services had crippled his ability to meet new obligations. The number of weather stations was not adequate to support aviation, even with the addition of Army and Navy stations. Limited personnel had forced cutbacks in services, and demands by the insurance industry for timely, accurate data were taking a heavy toll. Chief Charles Marvin had explained this as follows: "In general terms, the Weather Bureau is suffering from the ravages of the war and the consequences of an enormous change in economic conditions. Its work is conducted under strained conditions by faithful personnel, largely discouraged by the slow and inadequate adjustment of Federal compensations to existing conditions of life." During the 1920s, increasing numbers of employees were leaving the bureau—some after 30 years—because their salaries did not support their families. Annually, 100 percent of the lower civil service grades turned over: the bureau was training meteorological observers who then left for better-paying jobs. Meteorologists with a bachelor's degree working for the Army Signal Service started at more than $2,500 per year, while Weather Bureau meteorologists (with master's degrees and 10 years of experience) only earned $1,800 per year—less than most shop employees earned at the Bureau of Standards or than most clerks received at the Department of Agriculture's Office of Experiment Stations. Nor was the salary discrepancy between the Weather Bureau and other science-based agencies limited to the lowest levels. In 1921, the Weather Bureau's chief was paid $5,000 to lead an organization with more than 200 stations and a budget of $2 million. While the chief of the Office of Experiment Stations received the same salary, he supervised an organization with only five stations and a budget of $250,000. The chief of the Bureau of Chemistry and Soils was paid $8,000. His organization's budget was only $1.3 million.[60] The Weather Bureau was unique in this discrepancy between wages paid and corresponding levels of responsibility. No wonder the Weather Bureau, as an organization, carried itself in the manner of one who has been constantly put upon. Weather Bureau employees *were* put upon. They were given neither the respect nor the corresponding remuneration accorded to employees of other scientific agencies.

In 1924, the Civil Service reclassified positions to align the pay scales for similar positions across government agencies, but Weather Bureau employees were still paid inadequately (considering level of education and training, responsibilities, and length of service) relative to other civil servants.

By the early 1930s, the Weather Bureau was seeking additional employees to provide aviation services. However, it had a difficult time finding enough trained men (there were no women in the bureau). Senior grades required degrees in mathematics or physics, preferably with some meteorology courses. However, so few colleges offered separate meteorology courses that the bureau could not make them a requirement for employment. All positions, regardless of educational background, were filled by competitive civil service examination.[61]

The personnel situation had deteriorated further once the government instituted Depression-era economy measures. In mid 1932, men over the age of 70 (with a few exceptions) were immediately retired. With this cut the Weather Bureau lost 25 of its most senior people, including two-thirds of those with earned PhDs. Of the latter, only three (Chief Charles Marvin, a meteorological physicist, and the head of the New Orleans field office) kept their jobs. Most of those retired had been heading field stations—a position for which the most important indicator of probable success was years of experience. Those remaining within the system lacked equivalent education and training. This situation adversely affected the bureau's ability to provide effective weather services.[62]

In 1933, the Weather Bureau endured more funding reductions and personnel losses. Congress appropriated $400,000 less for fiscal year 1934 than for 1933. It then imposed a spending limit that was an additional $800,000 below the appropriation—a total loss of $1.2 million. The final budget was just shy of $3 million. As a result, the bureau laid off 500 employees and closed more than twenty first-order stations (including those at Fort Worth, St. Paul, and San Jose) and a large number of substations. A number of departments (particularly Agriculture and Commerce) lost weather services due to the budget cuts. Worse yet, the bureau lost additional senior personnel. Many of those with 30 or more years of service were involuntarily retired. Some of the remaining employees kept their jobs by moving into lower-ranking positions. Along with everyone else in government, they took a 15 percent pay cut—not an incentive for recruiting younger meteorologists.[63]

Weather Bureau employees thus had many concerns during this period: loss of jobs, pay, and funding for goods and services, and intense pressure to maintain consistent, high-quality weather services. Therefore, it had been especially irksome to receive complaints from entities such as the American Society of Civil Engineers that Weather Bureau data were difficult to use. An ASCE committee looking into the bureau's methods found personnel to have an "inferiority complex," enhanced by weaker educational backgrounds and inadequate equipment for scientific investigations, compared to those working elsewhere in the Department of Agriculture. Oliver Fassig (the former chief of the climatology division, who had been involuntarily retired) fired back that the Weather Bureau did not exist to support special-interest groups. Furthermore, the bureau was still hampered by the attitude toward meteorology as a science that existed during its establishment in the late 1800s, i.e., it was not a "real science" like physics or chemistry. As a result, it suffered from long-standing "poor intellectual visibility."[64]

In 1935, the bureau began to climb out of this desperate situation when it hired three young meteorologists with newly earned MIT PhDs: Horace Byers, Harry Wexler, and Stephen Lichtblau. Their mission was to bring the Norwegian polar front and air mass theory—a theory recommended by the Science Advisory Board—to the bureau. Their mandate was to study how Bergen School techniques could be applied to North American weather, and then teach them to field meteorologists. But the addition of these three young men did not markedly improve the bureau's educational profile. Surveying personnel in the late 1930s, Byers found that only 27 percent of "professional personnel" had college degrees, and that half of those degrees were in science or engineering. By 1939, there were only five Weather Bureau employees with meteorology degrees.[65] This dearth of professionally educated meteorologists was largely due to the low opinion held by academe about meteorology as a scientific discipline before World War II.

Stagnant and then dwindling appropriations kept the Weather Bureau in a rut. Reduced funding exacerbated already low salary levels and the bureau's inability to expand observation stations. With no research budget, the bureau was not able to analyze the data it had collected from its 5,000 volunteer observers. With little congressional support, it had not sought out new analysis and forecasting methods until prodded by the Science Advisory Board. As the 1930s closed, long-range forecasting and mathematics-based objective forecasting techniques appeared to be in the distant future.

Looking Abroad for Inspiration: The Navy Aerological Service

Like the Weather Bureau, the Navy Aerological Service (Navy Aerology for short) had long operated with limited funds and manpower and with virtually no support from the battleship admirals leading the Navy. But unlike the Weather Bureau, the Navy had looked to the Bergen School and adopted its methods—a decision that would inform, shape, and strengthen the professional relationships between several major figures who would eventually influence the development of numerical weather prediction in the United States.

Having transferred its marine meteorological service (minus the weather information plotted on pilot charts) to the Weather Bureau in 1904, the Navy had paid scant attention to meteorological services until World War I. Then, demands from aviation units forced it to expand its meteorological mission. After the war, as discussed above, the bureau had resumed its role as the nation's sole provider of weather services. However, it soon became obvious that the bureau was not going to have an office near every one of the naval activities that were scattered along the Atlantic, Gulf, and Pacific coasts. "It is fully recognized," Secretary of Agriculture David Franklin Houston wrote to Secretary of the Navy Josephus Daniels on 14 January 1920, "that certain meteorological work and observations must of necessity be conducted by the Navy in connection with its operations at base stations and on vessels at sea, but such work does not involve duplication of effort. In fact, stations so maintained by the Navy will supplement those of the Weather Bureau and be valuable to it."[66]

The Navy's aerological mission would be to provide "detailed weather information to naval aviators and aeronauts" and to provide local weather forecasts when a Weather Bureau office was not close by.[67] That did not seem too onerous a task; however, with only five officers and two enlisted men remaining from war service, the Navy was far from able to meet all requests for meteorological support.

Because almost all weather observing and forecasting tasks had been absorbed by the Weather Bureau in the earliest days of the twentieth century, the Navy was not prepared to fill a rapidly expanding need for meteorological support. It had no meteorological specialists and only a few basic instruments. Naval air stations were very interested in obtaining "allowances" for meteorological equipment and personnel. While an "allowance" would not guarantee equipment and personnel, without an allowance naval air stations would never get them. However, sailors were unfamiliar with meteorology, and civilian meteorologists were unfamiliar with the Navy, so it took some time to put a new Navy meteorological organization together. Starting in 1917, Alexander G. McAdie, director of the Harvard-affiliated Blue Hill Meteorological Observatory, began to provide meteorological training to officers in conjunction with MIT's aviation ground school. At the request of Assistant Secretary of the Navy Franklin D. Roosevelt, McAdie accepted a reserve commission as a lieutenant commander in January 1918 and began to determine the Navy's aerological needs and organize an aerological service. Shortly thereafter, enlisted personnel started receiving meteorological training at Pelham Bay, New York. The Navy shipped 9 officers and 15 enlisted men to England for further training with the British Meteorological Office and then on to European assignments for the war's duration. By war's end, 50 officers and 200 enlisted men were providing meteorological services to a variety of naval activities.[68]

When the Navy determined that it had to accept responsibility for most of its own weather support, it was obvious that the seven remaining meteorological personnel could not fulfill the mission. Naval activities at home and abroad, as well as afloat units, needed weather forecasts. Because weather conditions were important to flight safety, the Navy established a meteorology school at Naval Air Station Pensacola, home of the flight school. Training for both officers and enlisted men covered the science of meteorology and its applications to naval operations. Enlisted men took a four-month course that prepared them for assignments at naval air stations, on aircraft tenders (ships that provided repairs and maintenance to seaplanes), and with other ships and stations. In addition to this in-house instruction, Weather Bureau headquarters provided some naval officers with two additional months of "post-graduate study." Of the six officers who graduated from the basic course, three went directly to field assignments, while the other three moved to Washington for further training. This advanced course included non-instrumental observations of weather (e.g., the significance of particular cloud types), discussions of flying weather, weather map construction, discussion and forecasts, and physics of the air. The bureau also gave the visiting officers free access to its library.[69]

Despite the Navy's laudable effort to establish a training program that would boost its numbers of meteorologically trained personnel, any naval officer who planned to maintain a successful naval career had to spend a considerable amount of time at sea or serving with the nascent aviation units. Consequently, receiving meteorological training was not high on the list of desirable career options. With insufficient volunteers, the Navy ordered officers who had little or no interest in meteorology to the training courses. They stayed within the aerological program for the minimum required time before transferring to more career-enhancing positions. These uninterested officers tended to lead inefficient weather stations, which contributed to weather-related aircraft incidents. In the worst of these accidents, the rigid airship *Shenandoah* went down in a line squall on 3 September 1925, killing

14 people. Coming just one day after the disappearance of two Pacific-based PN-9 seaplanes in the Pacific, the case for the necessity of good weather support had been made.[70]

The Navy needed to take a different approach to maintain a cadre of highly trained meteorologists who could apply their knowledge to naval operations. The difficulty: line officers, i.e., "warfighters" serving afloat, already considered themselves to be good weather forecasters. They spent their lives at sea and had to be able to read the skies for indications of future weather conditions. They felt no need for advanced training. Furthermore, remaining in a specialty area like aerology would have effectively ended their careers. Promotions depended on filling shipboard "combat" positions. Making weather forecasts to aid the fleet was not sufficient to guarantee advancement. Despite the training program in Pensacola, by 1925 there were only two naval officers practicing meteorology. One of them—Reichelderfer—was eventually destined to become the Chief of the Weather Bureau.[71]

Francis Wilton Reichelderfer had graduated from Northwestern University with a degree in chemistry in 1917, just as the United States was entering the Great War. Joining the naval reserve intending to become a pilot, he signed up for meteorology training and was assigned to Alexander McAdie's training unit at Blue Hill. Reichelderfer did earn his wings after the war was over, but he remained in the meteorological field. By 1922, he was the head of Navy Aerology (a position he held until 1928) and occupied the Navy's "desk" at the Weather Bureau's headquarters, where he filled a liaison function while pursuing his own studies of Bergen School techniques. With the demand for aviation forecasting increasing as the numbers of meteorological practitioners dwindled, Reichelderfer decided that the only solution was to establish a post-graduate course for Navy meteorologists. In 1926, Reichelderfer (by then a lieutenant commander) and Edward P. Warner (an MIT professor of aeronautical engineering who was serving as Assistant Secretary of the Navy for Aeronautics) established a two-year post-graduate course in meteorology. Reichelderfer argued that the importance of weather information for aviation missions was new and distinct from the previous use of forecasting to ensure safety at sea. The Weather Bureau took care of marine forecasts. The aviators needed special weather information (e.g., on cloud layers, fog, and strong winds) to make decisions on launching aviation missions that could include scouting and bombing. Because that kind of detailed information could not be transmitted via teletype, an officer needed to be on site to provide "over-the-counter" briefings and to answer questions.[72]

The Navy taught the first year of this new course, emphasizing advanced physics and mathematics, at the Naval Postgraduate School on the US Naval

Academy's campus in Annapolis. The second year, concentrating on meteorology, had to be taught elsewhere. Reichelderfer approached climatologist and eugenics proponent Robert DeCourcy Ward of Harvard about the possibility of hosting the course. Ward agreed to host it for one year, if MIT's physics and mathematics faculty would teach dynamic meteorology (which deals with the solution of hydrodynamical and thermodynamical equations as related to the full range of atmospheric motion). At the end of the first year, neither MIT nor Harvard had the faculty to carry out the Navy's proposed instructional program. However, MIT's Warner had convinced the Daniel Guggenheim Fund for the Promotion of Aeronautics that support for aeronautics meant more than research on aircraft design and construction. Meteorological instruction and research leading to more accurate forecasts were essential for safe flight. The Guggenheim Fund gave MIT $34,000 to fund the first 3 years of a meteorology course, and provided Carl-Gustav Rossby to lead it.[73]

The Swedish-born Rossby would in time emerge as the most influential theoretical meteorologist of the middle years of the twentieth century. He had studied mathematics, mechanics, and astronomy at the University of Stockholm before moving on to work with Vilhelm Bjerknes at the Geophysical Institute in Bergen. After two years there, he studied hydrodynamics at the University of Leipzig's Geophysical Institute. Returning to Sweden in 1921, he took a position with the Swedish Meteorological and Hydrological Institute while he completed his *filosofie licenciat* in mathematical physics at the University of Stockholm. Awarded a fellowship by the American-Scandinavian Foundation to study in the United States, the energetic, hard-driving Rossby joined the Weather Bureau's headquarters staff. While working on questions related to atmospheric turbulence, he attempted to persuade bureau forecasters to use Bergen School techniques. Weather Bureau meteorologists were not receptive, but Francis Reichelderfer was. A friendship blossomed, and this pair of meteorologists would continue to work together to advance the discipline until Rossby's death in 1957. Rossby, having irritated the Weather Bureau hierarchy and needing another position, was invited by the Guggenheim Fund to organize weather services for its model airway being constructed between Los Angeles and San Francisco in 1928. Once the weather services were turned over to the bureau, Rossby was available to lead the new MIT meteorology program.[74]

Rossby established the course at MIT (within the department of aeronautical engineering) with the help of synoptic meteorologist Hurd C. Willett. (Synoptic meteorology is the subdiscipline that coordinates observations into a picture of the day's weather and makes predictions of future weather.) Willett had joined the Weather Bureau after graduating from Princeton in

1924, and subsequently spent time studying with the Bergen School. He was completing his PhD at George Washington University when he joined the new MIT program. The new curriculum included course work in physics of the air, and mathematical and dynamical meteorology, and practical work in forecasting. Reichelderfer hoped that the course would "arouse more general interest throughout the country in instruction in weather science and [lead] to fruitful research and development."[75]

MIT's new graduate program, based on Bergen School techniques, provided the Navy with a cadre of formally trained meteorologists. By 1934, 24 officers had attended and were working as aerologists. However, there was a continued lack of upward career mobility, and in 1940 only 18 would remain.[76] Thus, once again, the Navy would enter a war without sufficient personnel to provide the required meteorological support to the operating forces.

With so few meteorologists and with no research budget, the Navy aerologists, like their Weather Bureau counterparts, had little opportunity to implement new ideas and techniques. Despite these difficulties, Reichelderfer had circulated Jacob Bjerknes's first paper on frontal analysis techniques to his fellow Navy officers by 1921 and started applying those techniques to surface weather maps shortly thereafter. He actively sought papers written by Bergen School members and distributed them to colleagues. Therefore, Navy officers attending graduate school were familiar with the Norwegian methods. The Norwegian methods were also taught to Navy aerographer's mates (enlisted men) at the Aerology Observatory in Lakehurst, New Jersey, site of the airship base. While Reichelderfer was in Lakehurst to forecast for airship operations, he had Rossby's MIT group mail their daily weather maps to him. Even though the aerographer's mates were being trained in Bergen School methods, Reichelderfer noticed that their maps did not match the MIT maps. It was obvious to Reichelderfer that to successfully train the aerographer's mates in Bergen School techniques he needed to go to Norway.

En route to Norway in 1931, Reichelderfer spent almost a month with the British Meteorological Office to examine their organization and forecasting methods. A six-month stay in Bergen followed. Reichelderfer also visited weather offices all over Europe (including France and Germany), writing enthusiastically detailed accounts of their operations. He sent these reports (marked "Restricted") via diplomatic pouch from the US Embassy in Paris under naval intelligence cover sheets. Upon Reichelderfer's return, one of his new Norwegian colleagues traveled to the United States to lecture Navy aerologists. This led to what Reichelderfer later termed "successive invitations by universities that led to permanent residences by some of the well-known and distinguished Viking scientists."[77] Thus, the efforts of both Rossby and

Reichelderfer to promote Bergen School methods significantly influenced the eventual immigration of Scandinavian meteorologists to the United States. This influx of Scandinavian expertise would have a tremendous impact on the advancement of meteorology in America.

Mid-1930s newspaper articles typically described this "new" air mass analysis method to be of recent American origin, though in fact it was a Norwegian import. Infrequent weather observations, coupled with inadequate spatial distribution, impeded its full implementation.[78] To be effective, weather observations had to be taken nationwide every six hours, and data density had to increase. This was not a small concern for the Navy, which obtained all weather data from the Weather Bureau, whose budget could barely handle current requirements.

The Navy was also actively encouraging (and carrying out) the collection of upper-air observations. In the mid 1930s, Navy aerologists became the first to use special recording instruments (meteorographs) attached to airplanes to obtain temperature, pressure, and humidity data, which could then be used for local area forecasting and to supplement Norwegian methods. Navy Aerology was committed to staying current with the latest scientific developments—coming primarily from overseas—so as to advance meteorology in the United States and stand ready to fulfill its duties in war and peace. Weather forecasts for flight operations, visibility forecasts for the accurate firing of shipboard guns, and wind forecasts for ballistic targeting would all be important as the Navy prepared to enter another war in the late 1930s.[79]

Fighting to "Ground" Meteorology: The Army Signal Corps

The Army Signal Corps (Meteorological Division) had a longer history than either the Weather Bureau or the Navy's aerological service. Weather services in the United States had been a function of the Army Signal Corps from 1870 until their transfer to the Department of Agriculture in 1891.[80] The War Department then depended on the Weather Bureau for meteorological support until World War I.

Unlike earlier armed conflicts, the Army had recognized that weather support would be crucial to its success on the battlefield as it prepared to enter the war. Weather services had not become important because predictions were significantly better, or because Army leaders had determined that weather conditions affected battles. Rather, weather prediction mattered because advances in armaments dictated requirements for meteorological support. Artillery ranges had increased to ten miles or more; atmospheric conditions, winds in particular, influenced targeting. Army units were also using listening posts to determine the location of enemy artillery batteries. Known as "sound ranging," the accuracy of this woefully inadequate method decreased dramatically without knowledge of air density, and wind speed and direction. The successful use of poison gas depended on favorable winds. High winds blew the gas over enemy trenches or dispersed it too rapidly to be effective. Light winds carried the gas so slowly that the enemy could take countermeasures. If the wind shifted, it would drift back over friendly forces—an event that, as one contemporary observer noted, "seriously interfere[d] with the career of the gas officer." Therefore, accurate knowledge of the wind regime was very important. And, of course, the introduction of aviation assets meant flight forecasts. The Air Service of the American Expeditionary Force was one of the first Army organizations to require weather support. These early aviation forecasts had been for safety, not tactics. Their purpose was to keep these planes built from "wood, glue, wire, and fabric" out of adverse weather—high winds, turbulence, and hailstorms—that could bring them down.[81]

General John Joseph "Black Jack" Pershing, Commander of the American Expeditionary Force, had requested meteorological personnel. Like the Navy, the Army did not have enough meteorological officers to meet the demand. Not only did Pershing need meteorologists in wartime Europe; the Weather Bureau could not support the Army's stateside activities—the Gas Warfare Service, ordnance proving grounds, and field and coast artillery units.[82]

Manpower, not surprisingly, was hard to find. The Chief Signal Officer, General G. O. Squier, had called on the National Research Council to recommend possible sources of potential officers. Squier also had asked the Weather Bureau for help because "virtually all the trained meteorologists in the country were employed by the [bureau]." A planning committee composed of bureau personnel was led by Lieutenant Colonel Robert Millikan, then serving as the Officer in Charge of the Signal Corps Science Research Division. The committee determined that available assets had to be divided among three basic support areas: the American Expeditionary Force, the stateside activities needing weather services, and research into meteorological topics. To solve the manpower shortfall, the bureau had donated 25 percent of its 600 employees to the Army through the end of World War I. Hundreds more were trained—some at Texas A&M (figure 1.6)—just for wartime military service.[83]

The Signal Corps' provision of weather services was one part science, one part military tactics. Meteorology personnel faced challenges that fell into one of three categories. First, they had to develop statistical meteorology (i.e., clima-

Figure 1.6
Army Signal Corps meteorology training during World War I. These young trainees at Texas A&M are watching the launch of a weather balloon. (courtesy of Military Heritage Institute, Carlisle Barracks, Pennsylvania)

tology) to determine appropriate locations for military units and aerodromes. Second, they had to provide current meteorological information to military units, including ballistic winds to artillery companies, humidity, temperature, and wind data for sound ranging, and pilot-ballon and theodolite observations for aviation. Third, they had to provide forecasts in advance of military operations. Observers had to measure temperature, air density, and wind direction and speed so artillery units could exploit sound ranging and properly aim their guns.[84]

Cobbled together just as the United States entered the war, the meteorological division perform ed well during the conflict. Unfortunately, like their Navy brethren, meteorological personnel had left the Army in droves and returned to their peacetime occupations after the war. However, the mission remained. Planes were still flying. The Chemical Warfare Service had continued to conduct experiments and practice maneuvers. The field artillery units still needed standard ballistic range tables for their artillery pieces.[85]

Despite the hundreds of men trained in meteorology in World War I, between 1921 and 1935 no more than eleven weather officers served in the Signal Corps. Along with a handful of enlisted men, they were able to fulfill less than one-fifth of the demand for their services. The Signal Corps continued to build more weather stations (they quadrupled the number to approximately 40), but with so few soldiers the quality of meteorological services remained poor.

The Signal Corps trained both officers and enlisted weather personnel at Camp Vail (later Fort Monmouth), New Jersey, which also was home to meteorological instrument development. Additional enlisted men received meteorology training at Carlstrom Field (Florida) and March Field (California) as part of flight training. With close ties to the Weather Bureau, which was providing most of the forecasts, the Signal Corps took no interest in the Bergen School techniques. When Reichelderfer offered Signal Corps leaders the chance to participate in the Navy's new MIT graduate program, they declined.[86]

The Air Service had not been content waiting for the Signal Corps to upgrade weather support. From 1922 to 1924, the meteorology section's budget more than doubled from $27,000 to $67,000—and it was all due to Air Service requirements. As far as the Air Service was concerned, meteorological services belonged under its jurisdiction. The Signal Corps argued that weather services were not exclusive to the Air Corps. Therefore, the meteorology section stayed within the Signal Corps, but fell under the Intelligence Division, having escaped from the Special Services Division that supervised the Pigeon, Photo, and Commercial sections.[87]

From the mid 1920s on, a power struggle ensued between the older Army "ground pounders" and the younger aviators. The latter wanted more support. The former had control and intended to retain it. The Signal Corps did not care much about the Meteorology Section, and probably could have been forced to give it up. However, the Air Service was fighting for survival within the military structure and did not have the time or the energy to become embroiled over what seemed like a minor point.[88] Although the Air Service was not effectively fighting for control over meteorology, it did accept Reichelderfer's offer and sent its first student to MIT in fall 1929. However, Air Service meteorologists did not embrace Bergen School methods until 1935.[89]

In 1934, the Signal Corps' chief declared that he wanted release from the weather mission if he did not get more funding. But that same year, charges that the US Post Office had fraudulently awarded air mail routes without competitive bidding (later shown to be unwarranted) led Post Office Solicitor Karl Crowley to declare that existing contracts were void. President Roosevelt then issued Executive Order 6591, which canceled all air mail contracts and ordered the Army to carry the mail. The Army Air Service took over the flights with only ten days' notice. This was an ill-fated decision. Not only were Army aircraft inferior to anything being flown by commercial carriers, the weather was particularly bad and the forecasts were inadequate. Within three weeks, twelve pilots died in airplane crashes. Although this decision became a political liability for Roosevelt, the Signal Corps bore the brunt of criticism for its forecasting deficiencies. The Signal Corps' gambit to use this fiasco to obtain additional funds did not work, however, and the Air Corps, which had more weather-trained officers than the Signal Corps, took over weather forecasting responsibilities as the primary user of the services.[90] Commenting on the divorce of meteorology from the Signal Corps, one officer later testified that meteorology had "no more to do with signals than Donald Duck."[91]

In 1937, the Air Corps began sponsoring weather services for aviation and for ground forces serving in units at the division level or above. Signal Corps weather officers desiring transfers had to qualify as pilots—a requirement that did not encourage movement of trained personnel. Enough personnel were attracted to this new meteorological service that by the end of 1939 30 officers and almost 400 enlisted men were serving in the Air Corps.[92] Even this increase would not be nearly enough to provide for the requirements of the by then rapidly approaching war.

While the forecasting mission moved to the Air Corps, the limited meteorological research function remained within the Signal Corps. Most research and development activities were centered on meteorological instruments. Despite pressure to move this work to Wright Field in Ohio due to aviation requirements, it remained at Camp Vail. The Army needed meteorological support for all of its forces, not just the aviators.

The work at Fort Monmouth later proved critical for the eventual development of numerical weather prediction models. The researchers worked to develop an audiomodulated radiosonde (then called a radiometeorograph). This instrument would allow upper-air observers to gather data during the night or during cloudy weather—whenever a pilot balloon would normally be obscured. Large balloons carrying meteorological instruments and a radio transmitter would send data to a receiving station, which was far superior than trying to find the recording equipment after it had fallen back to the ground. The Army also conducted meteorological research related to chemical warfare. The Chemical Warfare Service sponsored almost 700 projects for the Army, the Navy, and civilian organizations. However, appropriations were so small (less than $1 million annually for all projects combined from 1923 to 1926, and less than $2 million annually from 1927 to 1938) that each project received an average of 2–3 thousand dollars. Most research was directed, not to basic weather research or prediction, but toward the effects of micrometeorological phenomena on the movement of gas.[93]

In Retrospect: Weather Services in the Interwar Period

While European weather services were relatively awash with money in the period between the two world wars, encouraged research, and eagerly tried out the new ideas of the Bergen School, the United States experienced a period of retrenchment for all of the weather services. For example, while the Norwegians and the Germans funded geophysical institutes to conduct meteorological research, the Americans did not. The military services experienced dramatic drops in personnel and funding immediately after World War I, from which they did not start to recover until war loomed once again. The Weather Bureau, forced to operate without almost a quarter of its personnel during World War I, got them back, only to face stagnant appropriations in the 1920s, then drastically reduced appropriations as the Great Depression deepened.

Losing its most experienced personnel, and having no opportunity to replace them, the Weather Bureau could barely provide routine services, much less expanded aviation services. With low levels of funding and compensation, and with a training program that assumed that the only way to create a forecaster was through an apprenticeship of 5–6 years, it is hardly surprising that the bureau was not in the forefront of implementing new forecasting techniques. Furthermore, astrophysicists, statisticians, economists, physicists remote from atmospheric work, and others were telling the Weather Bureau how it should be doing its job. What could have been more demoralizing for government meteorologists than repeatedly being told, through words and deeds, that their scientific discipline did not deserve to be included with the scientific "big boys"?

The Navy Aerological Service had suffered more from benign neglect. Sea captains, believing they knew everything there was to know about the weather, were perfectly satisfied with their ability to operate under any conditions. For decades, they (and commanders of entire fleets) failed to consider the influence of weather on naval operations. However, operating in weather extremes was just part of fulfilling the Navy's mission at sea. This confident devil-may-care way of thinking did not last as long in the aviation community. The atmosphere is much less forgiving to aircraft under less than perfect conditions. Therefore, aviators demanded increased meteorological support even while the Navy did not provide a career path for those who would provide it.

The interwar period saw the Navy adopt and spread Bergen School methods through its own professional networks in a way that did not occur in the much larger Weather Bureau, still top-heavy with older men. Thus, when the Science Advisory Board directed the adoption of air mass analysis techniques, the Navy was able to meet the requirements with less resistance. Instruments were being developed and put on aircraft to gather data, which were then shared with the Weather Bureau. In this way, the Navy was looking forward. What it could not see was how once again it was approaching a time of war with insufficient personnel to fulfill its mission.

As World War II loomed, the Meteorology Section of the Army Signal Corps was probably in the worst shape of all the weather services. It was a low-priority organization, thrown together with the messenger pigeons. Aviation units received their forecasts from the Weather Bureau. With no war in sight for the United States through the 1920s and into the 1930s, there was little concern about providing meteorological services to ground troops overseas. Research was almost exclusively focused on developing and improving meteorological instruments. Although these newly developed instruments, in particular the prototype radiosonde, greatly enhanced meteorologists' ability to collect upper-air data, the success of that endeavor was not sufficient to keep the Signal Corps' Meteorological Section going. Scant attention was paid to new developments in the atmospheric sciences and the old methods—good enough for the Weather Bureau—were good enough for the Signal Service. It was not until the end of this period that the Air Corps prevailed and the meteorological mission was moved out of the Signal Corps. The focus then shifted to keeping aviation assets (pilots and aircraft) safe and effective.

Meteorological services advanced in very limited ways in the United States between the wars. Instrumentation improved primarily through the efforts of the Signal Corps' research arm and because of the interest of instrument specialist Charles Marvin, then Chief of the Weather Bureau. High profile criticism of the Weather Bureau had prompted Secretary of Agriculture Wallace to make the politically expedient move to call in outside "experts" in the guise of the Science Advisory Board to recommend ways to "fix" the bureau. However, many of the bureau's shortcomings are better attributed not to a failure of leadership, but to a failure of adequate funding for a government organization providing a free service that earned business and agriculture interests millions of dollars a year. Not even the distinguished members of the Science Advisory Board could secure the funding the bureau needed—it just recommended changes that the bureau could not afford.

And so America's weather services limped along, doing their best to ensure safety of flight, warn farmers and the general public of weather hazards, and get out a forecast that made sense. As the world marched steadily toward global conflict, the Weather Bureau, the Navy Aerological Service, and the Air Corps' new weather section would soon be put to a huge test—a test for which none of them was ready.

2 :: Toward a More Dynamic Atmosphere: Discipline Development in the Interwar Period (1919–1938)

The research university structure had emerged in the United States in the late nineteenth century. The physical and life sciences, and later the earth sciences, continued to develop as major disciplinary communities into the early twentieth century. A National Research Council compilation of doctoral degree data from 1923 illustrates the significant differences in disciplinary strength. From 1919 to 1923, more than 600 doctoral degrees were awarded in chemistry, approximately 200 each in botany and physics, almost 100 in the geological sciences, and 20 in astronomy. There were only *two* PhDs awarded in meteorology,[1] a discipline that had been of scientific interest for hundreds of years.

Writing in 1918, Harvard climatologist Robert DeCourcy Ward opined that meteorology was not more widely studied because having spent a lifetime becoming familiar with meteorological phenomena makes every man think himself a "born meteorologist." And that very familiarity, Ward continued, "breeds a certain degree of contempt." Because weather is a topic of everyday conversation, serious study did not seem to be worthwhile. Ward quoted a "highly educated" woman of his acquaintance: "You have a very difficult subject to teach. People, generally, do not care to hear about things which they think that they already know."[2] Everyone, in essence, *was* a meteorologist.

In addition to the general lack of interest due to "knowing it all," there were so few meteorologists that finding instructors was difficult. Ward hoped that faculty attached to physics, geography, geology, or other more marginally related science departments would take it upon themselves to learn enough meteorology to prepare and teach an elementary course. With their interest sufficiently piqued, students might demand more advanced courses, thus opening the door for the establishment of meteorology curricula in colleges across the country.[3] Ward's dream would eventually come true, but it would prove to be a very slow process.

The massive meteorological training effort that had produced military weather forecasters and observers during World War I spurred renewed interest in meteorological education within the civilian community. However, meteorology was not yet a university discipline; it was hardly a scientific discipline at all. Even before the 1920s, major universities across the United States already supported robust departments of physics, biology, chemistry, astronomy, and geology with well-organized graduate programs. Meteorology was seldom taught. For meteorology to become a respectable discipline recognized by other scientific communities, meteorologists needed to create the two essential elements of any professional discipline: dedicated academic programs in major research universities and a professional society, neither of which existed in the United States at war's end. But the impetus provided by the war led to the establishment of academic meteorology and a professional society. By the late 1930s, MIT, Caltech, and NYU had meteorology programs and the American Meteorological Society aggressively sought to push meteorology into the scientific mainstream. The professionalization of meteorology had begun.

Organizing Academics: From Military Needs to Civilian Wants

As World War I came to a close, meteorological instruction in the United States took one of three approaches: (1) the climatological approach espoused by Ward at Harvard, (2) the physical approach of the Weather Bureau's atmospheric physicist W. J. Humphreys, or (3) the combined climo-physical approach of Charles F. Brooks of the Blue Hill Meteorological Observatory. The latter had been the approach used at the Army Signal Service meteorology school, which had trained more people in meteorology in a shorter period of time than any other organization in the United States.[4]

Planning for the Signal Corps school had started in fall 1917. The Army had needed to quickly train 1,000 men in meteorology to serve as military weather observers and forecasters. As the United States entered the war, the only trained weather observers were those working for the Weather Bureau. They had already been inducted into the military, and more were needed.[5]

The bureau trained these new recruits within its own offices until becoming overwhelmed by the sheer numbers of soldiers. By spring 1918, the Signal Corps had established at Texas A&M University a special school that included instruction on the physical properties of the atmosphere, weather forecasting, the different uses of meteorology (aeronautics, agriculture, commerce), and the physiological effects of weather and weather changes.[6] This stands in stark contrast to the *mobilization* of chemists and physicists during the same conflict. Those disciplines possessed ample numbers of scientists who just needed to be brought in under the military umbrella—usually as reserve officers.[7] Meteorologists, both professionals and sub-professionals, had to be trained.

Robert Millikan—organizer of the entire Army meteorology training program—required that all weather observer school recruits have college degrees, preferably in mathematics, science, or engineering. This was a startling requirement, considering that by the start of the twenty-first century—despite all the highly technical equipment involved—observers generally possessed only high school diplomas. The engineers, who comprised more than half of the original 300 trainees, were the most interested in aerological work. Their education and training enabled them to suggest new designs for meteorological instruments and to develop faster methods for computing ballistic wind values needed by artillery units.[8]

Academic and Weather Bureau meteorologists knew that military training was not equivalent to the kinds of college courses that would be necessary to create a cadre of professional meteorologists in the United States. As prominent British meteorologist Sir Napier Shaw argued, professional training was an absolute requirement for the advancement of the science. "Observations, map making, and forecasting don't a science make," Shaw argued; empirical knowledge was important, but would not by itself lead to knowledge of the atmosphere's physical processes.[9] To determine how those physical processes worked would take considerably more meteorologists educated in physics and mathematics.

What the academic and applied meteorologists did not know was the extent of existing meteorological instruction in American colleges and universities: course lengths, material covered, types of students reached, and who was offering them. During this period, at least three different researchers and organizations initiated studies of meteorological instruction. In 1919, the nation's largest employer of meteorologists—the Weather Bureau—asked the US Bureau of Education to survey higher education institutions throughout the country about their meteorological offerings. The Education Bureau sent out more than 600 questionnaires, and received almost 70 percent back. The Education Bureau assumed the unreturned questionnaires had gone to colleges with no offerings. Of those reporting, 84 percent of the colleges did not offer a separate meteorology course. Of these colleges, 22 percent offered meteorology as part of a more general course, while another 13 percent intended to offer it as a separate course. Ten colleges listed Weather Bureau employees as their instructors. Of the 70 colleges that did offer meteorology courses, 20 did so in the geology department, 10 in the physics department, and one each in the chemistry, biology, and astronomy departments. The remaining 53 percent did not specify which department offered their meteorology course.[10] Clearly, meteorology garnered limited attention in academe. As Caltech physicist and aerodynamicist Theodore von Kármán put it years later, "few academicians accepted meteorology because it was regarded as a guessing science."[11]

In 1920, leaders of the newly formed American Meteorological Society were concerned about the dearth of meteorological instruction. They formed the Committee on Meteorological Instruction to address this deficiency. Committee members concentrated their efforts on three areas: (1) collecting and publishing teaching techniques in the society's *Bulletin* (*BAMS*), which would improve meteorological instruction; (2) reviewing books that could be used by meteorology and/or climatology instructors and identifying other instructional sources; and (3) promoting the establishment of meteorology courses in colleges and universities where they were not being offered.[12] Owing to its importance to agriculture, at least one AMS member argued that meteorology certainly belonged at land grant institutions.[13] Cornell University and the Utah Agricultural College—both land grant schools—were already offering meteorology courses.

Poor career prospects exacerbated disciplinary development. Salaries were low, and before World War I there had been very little demand for meteorologists. During the war, demand far outstripped supply and triggered hastily assembled training programs. With the end of the war, expanding aeronautics interests in the civilian sector kept the demand higher than the supply, but years of poor career opportunities had left the profession without enough qualified people to teach the next generation. There were more openings for meteorology instructors than people to fill them.[14]

Even those courses that were offered sometimes provided minimal instruction. One example is a course in "Applied Meteorology" (meeting for less than an hour per week) taught at the Southern Branch of the University of California (forerunner of UCLA). In spring 1920, it gave students an overview of weather studies, including climatology, and how such information could be applied to commerce, agriculture and horticulture. The course's prerequisites included one year of physics, good algebra skills, and knowledge of general physiography. Students' grades were based on a couple of 300-word papers on such topics as "Advantages of Meteorology Study" and "High Pressure and Low Pressure Areas Compared." Given the requirements for the course, it is unclear why physics and algebra were high on the list of prerequisites. While their Scandinavian counterparts were doing air mass analysis, the Southern Californians were concerned with "Factors in a Healthful Climate."[15] Of the 84 students (mostly seniors), a few were engineers, some were in political science, and many were teachers taking the course to supplement their geography studies. A "detail" from the nearby Army Air Service balloon school also attended the course.[16] None of these students was likely to go further in meteorology, and it is questionable how many of the pre-service teachers received enough information to create an exciting approach to studying the weather in their classrooms.

Before recommendations could be made on the improvement of meteorological instruction, the AMS committee thought it would be important to determine the status of such instruction at all levels of education—primary, secondary, and post-secondary. Reporting their findings at the first AMS annual meeting in 1920, the committee members noted that while meteorology was of great economic importance to the country and was an equally important part of a liberal education, it was not accorded an important position in the educational system. The need for such instruction was widely recognized: primary schools had nature studies, for example, and high schools taught physiography, so there were already niches that could absorb meteorology. Unfortunately, the lack of meteorology courses at teacher training institutions translated into ill-prepared primary and secondary teachers. The committee concluded that once higher education institutions offered meteorology it often became a very popular course for liberal arts and education students. The report, while not directly addressing the lack of a more advanced meteorology curriculum in the country, did encourage the AMS to get involved in establishing such a program to increase the pool of qualified professional meteorologists.[17]

By the next year, more colleges were adding meteorology to their curricula. The most ambitious program, and the one that produced the most graduate students during the 1920s, was Clark University's newly established geography school, which also offered meteorology and climatology courses.[18] But movement was still too slow for the AMS Committee on Meteorological Instruction. In 1922, the members proposed creating and sending "propaganda . . . often and with emphasis" to the head of every American college and university, extolling the benefits of meteorology and climatology courses. The departmental targets of choice: physics, geology, geography, and astronomy.[19]

Despite the lamentations, meteorology and climatology instruction at some of America's more distinguished schools was on the upswing in the 1920s. Harvard students were allowed to "concentrate" in meteorology and climatology (a curriculum which entailed six courses including mathematics and physics) for undergraduates and a research course in climatology and aerology for graduate students. Cornell reported an overflow of applicants for its meteorology course, which was limited to 100 students.[20]

A survey of normal school and teacher's college catalogs in 1928 showed that 35 courses were being offered in basic meteorology or climatology and its effects on man; 150 focused on geographical attributes, including climatology, of specific continents or countries; and another 50 geography courses included elements of meteorology or climatology. Despite the overall increase in meteorology courses in teacher training institutions, meteorology had not yet appeared in the all-important methods courses, wherein students learned age-appropriate techniques for teaching different subjects.[21]

The most significant new courses—created at MIT at the behest of the Navy and financially supported by the Guggenheim Fund—started in the fall of 1928. Almost 10 years after the AMS started lobbying for advanced meteorological instruction in the United States, it appeared to be happening. Aviation interests—military and civilian—were primarily responsible. As the decade closed, meteorologists still saw the need for "truly serious training" in meteorology, grounded in physics and mathematics, as an "urgent" requirement. They hoped that the need would be met by the "most progressive institutions."[22] However, the lack of academic meteorologists meant colleges had to rely on US Weather Bureau and Canadian Meteorological Office meteorologists to fill instructor positions.[23]

Despite the catastrophic economic upheaval of the Great Depression, the AMS continued to pursue its goal of expanded educational opportunities in meteorology. In 1934, updated information about the state of meteorological education in the United States came from an unlikely quarter: the University of Southern California's School of Education. Graduate student Woodrow C. Jacobs, writing his master's thesis on meteorology instruction in American higher education, analyzed the breadth and depth of available meteorology courses. Receiving 733 college catalogs in response to his query letters, Jacobs followed up by sending a questionnaire to each school offering at least one meteorology course. He found that meteorology courses (those which were mostly meteorology, not just a subset of a course) had increased slowly and steadily until 1924. The burgeoning popular interest in aviation had precipitated a dramatic increase in courses starting in 1925. The greatest expansion had occurred at teacher's colleges and technical institutions. Indeed, over half of all such courses were added after 1924. Almost a quarter of the colleges offered meteorology courses, ranging from a high of twelve at Clark University to a low of one. State colleges and universities, as well as teacher training institutions, often offered meteorology courses, but they were rarely taught at private, technical/engineering, or women's colleges. Most were just one semester long.[24]

Found in twenty different academic departments, meteorology courses were more likely to be among the offerings of geography (46 percent) than physics departments (8 percent) even though meteorology was based on physical laws. Four colleges reported having meteorology departments. Most geography offerings were based on climatology or were of the "weather and man" variety of meteorology. "[The] study of meteorology as a pure or applied science," Woodrow Jacobs concluded, "seems to have been relegated to the background in most cases, a situation which is not generally true of science study in the colleges and universities of this country." Most instructors were well qualified in their own disciplines, but they had little or no training in meteorology. So while meteorology offerings were quite extensive, the quality of undergraduate courses, and graduate and research work, was poor.[25]

Courses offered at teacher training institutions were deliberately nontechnical and required scant physics. Most included discussions of the physical and chemical characteristics of air, as well as climatological information, and instruction in taking measurements of temperature, pressure, and humidity. Suitable textbooks were lacking. Faculty needed texts that provided a survey of meteorology that did not depend on physics.[26]

As the interwar period ended, only one other institution besides MIT established a meteorology department: New York University. Meteorology courses had been offered within its geology department for a number of years, but the separate department led by Rossby-trained South African Athelstan Spilhaus opened in 1937.[27] Its curriculum included a general undergraduate program designed to meet the needs of airlines, the Weather Bureau, and "other potential employers." The graduate program, however, was within the College of Engineering. Students took upper-air and surface observations—which were sent to the Weather Bureau—from the university's meteorological laboratory.[28]

Meteorology instruction expanded dramatically in the interwar period. However, it was dominated by a non-technical approach in the geography departments of teacher training institutions. Two decades after World War I had underscored the importance of meteorology to the nation's security, advanced instructional opportunities remained severely limited, particularly when compared with other scientific disciplines. The hoped for theoretical physics- and mathematics-based instruction that would lead to graduate study and research was available at only two institutions: MIT and NYU. And those meteorology programs did not have the resources to attempt major research projects. Most of the research that was conducted had limited value to forecasting applications.

The small number of academic meteorologists active during the interwar period spent much time getting their instructional programs started. However, graduate thesis topics provide a glimpse of the emerging research agenda. Doctoral programs in meteorology were non-existent in the immediate postwar years, so graduate students obtained their degrees through physics and geography departments. The former led to dynamical or physical topics, while the latter tended toward climatology.

Two doctoral degrees were awarded in 1922—a modest start—but 7 years passed before another four were awarded in 1929. The first two degrees—one from Cornell and the other from George Washington University—addressed climate and dynamics topics respectively. The latter, on free-air pressure maps, was written by aviation pioneer Clarence LeRoy Meisinger, whose promising meteorology career was cut short when lightning struck the balloon in which he was taking upper-air measurements.[29] Of the four degrees awarded at the end of the decade, two were climatology-based theses by geographers, but the other two (from George Washington University) were more meteorological. One, on the geometric theory of halos, went to Edgar W. Woolard (who opted to be a mathematics professor). The other went to Weather Bureau employee Hurd C. Willett for his studies of fog and haze. Upon receiving his degree, Willett left the Weather Bureau and joined Rossby on the MIT faculty.[30]

Since homegrown academic meteorologists were rare, meteorological research was understandably limited. A gradual influx of European meteorologists—which increased dramatically as Hitler expanded the Third Reich's reach throughout Europe and visiting scholars were trapped in the United States—boosted the numbers of researchers. Europeans were coming from universities and institutes where meteorology had the same standing as astronomy; it was considered a "real" science. Cutting-edge meteorological developments were emerging from the Bergen School and some of its trainees and practitioners were spending time in the United States. Bernhard Haurwitz of the University of Leipzig's Geophysical Institute served as a Research Fellow at MIT and Harvard's Blue Hill Meteorological Observatory during the 1932–33 academic year. The Swedish-born and Swedish-trained Rossby periodically returned to Scandinavia to keep up with new approaches and to advance his own research in dynamical meteorology.[31]

As the 1930s unfolded and meteorology programs took root at NYU and at Caltech, the research agendas at these institutions, in addition to those of MIT and Blue Hill, took on distinctive attributes. NYU concentrated on investigating the upper air and visibility, i.e., those topics of the most immediate importance to aeronautics, as they pertained to local New York conditions.[32]

MIT took a more geophysical approach, emphasizing the development and application of the Bergen School's polar front theories along with empirical work on the movement of air masses.[33] All of its more theoretical work depended on mathematical or quantitative approaches, but this did not preclude synoptic work that emphasized weather map analysis as an adjunct to explaining atmospheric phenomena as well as forecasting weather events.[34] MIT personnel also conducted research on the improvement of fog forecasting as a precursor to either preventing its development or dispersing it once formed at airports and landing strips.[35] Toward the end of this period, financial support from the Bankhead-Jones Fund (under the supervision of the US Department of Agriculture) enabled the beginnings of research into long-range weather forecasting and its implications for agriculture.[36] Caltech's program, under the direction of Irving Krick in the mid 1930s, concentrated on applied meteorology at the expense of theory and was primarily a training program for meteorologists intending to work for the airline industry and other applied activities. Krick, who had entered meteorology at the behest of his brother-in-law, Horace Byers, received his PhD at Caltech after struggling through the degree program's mathematical theory. He ingratiated himself with aviation and film industry executives in Southern California, and built up a substantial consulting business for long-range forecasting (based on "weather typing"—the matching of current weather patterns with ones that had occurred in the past) and weather modification while still on the Caltech staff.[37] The Blue Hill Meteorological Observatory (connected with, but not funded by, Harvard University) also conducted a variety of research projects during this period. Blue Hill researchers worked on instrumentation (particularly radiosonde development and deployment) for collecting upper-air data, on atmospheric phenomena of importance to aviators (thunderstorms, lightning, icing, and fog), on dust measurements, and on a variety of solar radiation measurement programs and associated data analysis.[38] However, there was no focused research program coordinated among the handful of groups doing theoretical work. Each took its own path, although MIT, NYU, and Caltech all were either funded by or closely associated with aviation interests—the driving force behind increasing interest in the science.

During the 1920s, the tightly woven prewar relationship between meteorology and climatology began to unravel. Geography departments settled on a climatology (both statistical and descriptive) model that would serve the needs of their economic and cultural subdisciplines. Teachers' colleges remained non-technical, providing no opportunity for meteorology as a physical science to take hold. The Navy's need for advanced instruction in meteorology, coupled with the financial backing of the Guggenheim Fund, enabled MIT to establish a mathematics- and physics-based meteorology curriculum under the auspices of its aeronautical engineering department. With the hiring of Bergen School acolyte Carl-Gustav Rossby by MIT, the door opened to spread the air mass and polar front theories among American meteorologists.

Research based on describing the atmosphere, which had tended to dominate during the years before the Great War, went into decline. In the postwar years, research became increasingly theoretical and focused on the dynamical and physical properties of the atmosphere. Development of instrumentation, while perhaps not seen directly as atmospheric research, was necessary to the furthering of the research agenda aimed at applying physical laws to atmospheric data. Scandinavian scientists visiting the United States brought new ideas and approaches that could no longer be ignored. The graduate students of the 1930s, steeped in Bergen School methods, would play a decisive role in spreading them across the country. With war—particularly a war that would be fought in the air—looming in the late 1930s, knowledge of the atmosphere would prove to be of the utmost importance to a successful military outcome. That success would, in turn, gain disciplinary respect for a rapidly growing community of professional meteorologists.

On the Path to Professionalization

As previously noted, the operational needs of the military services during World War I were responsible for a surge of interest in meteorological support and the training necessary to ensure its provision. Once the war was over, interest did not subside. This was true because increased aviation activity in both the civilian and military sectors as well as an awareness of the impact of weather conditions on commerce, agriculture, and health led to increasing demand for additional educational opportunities, more trained meteorologists, and a variety of specialized forecasts from the Weather Bureau. The immediate postwar period was an opportune time for the far-flung meteorology community to capitalize on this strong interest. Ultimately all these factors encouraged the organization of the American Meteorological Society (AMS) at the 1920 American Association for the Advancement of Science (AAAS) meeting in St. Louis. Its stated purpose was "the advancement and diffusion of knowledge of meteorology, including climatology, and the development of its application to public health, agriculture, engineering, transportation by land and inland waterways, navigation of the air and oceans, and other forms of industry and commerce."[39]

Considering that meteorology had been a topic of discussion since language had developed, and a part of natural philosophy since the Greeks, it is somewhat incongruous that no formal, broad-based society had arisen to represent the professional interests of meteorologists in the United States before 1919. By the time the American Meteorological Society was founded, the American Chemical Society (founded in 1876), the American Physical Society (1899), and the American Astronomical Society (1899) were all firmly established.

One might assume that the American Meteorological Society was the brainchild of either academic or Weather Bureau meteorologists. It was not. On the contrary, it sprang from the mind of a Signal Corps sergeant, Perez Etkes, a former student of Charles F. Brooks at the Signal Corps Meteorology School. In a letter to Brooks, Etkes argued that the aviators with whom he worked realized the importance of weather but did not have enough specific atmospheric information to find it valuable. Therefore, he proposed an "American Meteorological Institute" for the purpose of spreading meteorology "amongst the people" by establishing weather stations in schools and offering opportunities for graduate education via prizes and scholarships. Brooks was less than enthusiastic. However, he grudgingly acknowledged that it would not be that difficult to find more than 100 members for a meteorological organization from the ranks of the Signal Corps, the Weather Bureau, and meteorology educators. Even with limited dues, such a society could publish its own "periodical leaflet" containing items not normally published in *Monthly Weather Review*, which was then the only venue for publishing meteorological research. The new leaflet would, nonetheless, foster meteorological research by providing an outlet for members' ideas and professional concerns. The more Brooks thought about, the better the idea seemed.

Pursuing Etkes's proposed organization, Brooks further consulted with meteorologists within the Weather Bureau and academe. Because most meteorological work was being done by the Weather Bureau and since the Association of American Geographers already had a niche for climatology (and by association meteorology), Brooks originally considered an organization of meteorology instructors. That seemed too restrictive. A general organization of both professionals and amateurs, which allowed meteorologists to get together and talk, struck Brooks as the better approach. The primary stumbling block: the lack of meteorologists. William Morris Davis, professor emeritus of geology at Harvard, wrote to Brooks: "You can get a lot of men who dabble; and a lot of men who add up temperatures and divide by 30; but meteorologists are birds of a different feather."[40]

Despite Davis's warning, Brooks persisted. Brooks saw the future AMS as a way to promote badly needed meteorological instruction and research. The membership base he chose emphasized teachers, Weather Bureau employees, and current and former Signal Corps meteorologists and Navy aerologists. More than 900—significantly more than the original 100 Brooks thought could be enticed to join—persons equally divided between professionals and amateurs signed up the first year.[41]

To fulfill the AMS's mission of promoting research and instruction, members immediately formed eleven committees to address "the advancement and diffusion" of meteorology and "the development of numerous applications of meteorology to human affairs." The reports of topics discussed at the initial committee meetings give an indication of what were seen as the primary disciplinary goals in this early period.[42]

The Research, Meteorological Instruction, Public Information, and Membership committees composed the groups working on "advancement and diffusion." The Research Committee, with all but one member from the Weather Bureau, observed the lack of a strong independent group of meteorologists in the United States. Therefore, the committee's initial work would need to be "educational" to spark interest and direct subsequent research ideas down an appropriate path.[43] The Meteorological Instruction Committee, as discussed above, focused on expanding meteorological education throughout all levels of the school system.[44] The mission of the Public Information Committee was one that could be repeated by just about any scientific organization throughout the twentieth century: to eradicate popular errors concerning weather and meteorology and replace them with correct information by enlisting the aid of the newspapers and other media outlets. However, one of the "deeply rooted beliefs" that this group wanted to eradicate was that "the operations of mankind can have an important influence upon weather and climate."[45] In 1920, the idea that mankind's actions could affect weather and climate had not taken hold among professional meteorologists.

Committee members addressing meteorological applications established the following sections: physiology, agriculture, hydrology, business, commerce (transportation), marine, and aeronautical meteorology. With the exception of the physiology committee, the purposes of the others would remain self-evident into the twenty-first century. Their missions were to encourage meteorological research in areas that would be directly applicable to a given commercial segment and either add economic value or increase safety. The physiology committee's mission is perhaps not so obvious. Led by Yale geographer Ellsworth Huntington, committee members included representatives from medicine, sanitary engineering, and hydrography.[46] Their purpose was to bring meteorological information to a variety of disciplines concerned with the connection of weather conditions to health. To do this, the committee intended to promote sharing research results across disciplinary boundaries, and to teach physicians how to take and use simplified meteorological measurements to improve their patients' health. Members were concerned that not enough emphasis was put on establishing a link tying weather events and conditions to the general health of the population.[47]

No matter their primary mission, all AMS committees were assigned to "spread the word" about meteorology and climatology and their economic and cultural importance. Considering the sorry state of the discipline in the post-World War I United States, this would prove to be no small task.

Certainly American meteorology had had its moments of brilliance: the early theoretical work of nineteenth-century meteorologists James Pollard Espy, Cleveland Abbe, and William Ferrel, as well as the research-based Smithsonian meteorology project (1849–1874). However, as the pressure for weather services increased, the flame of weather research started to dim. Soon, the primary mission of meteorology, as embodied in the Weather Bureau, was to produce forecasts and warnings. Disciplinary advances would not, however, arise from the daily forecasting routine. To become a scientifically respected professional community, meteorology and its practitioners would need to develop an active research component.

For as long as anyone could remember, meteorology had first and foremost been involved with collecting data. Based on experience, those data were used to make forecasts. But the act of collecting does not make a science. If meteorology were going to move from art to science, then meteorologists had to apply mathematical and physical principles to the data. As George Washington University-trained mathematical meteorologist Edgar Woolard put it in 1923, the processes of weather are "simply examples of the operation of ordinary physical laws." He acknowledged that those laws would need some special treatment and despite all the collecting of data, the needed data (primarily a matter of spatial coverage and lack of upper-air reports) were just not there. He hoped that people with solid backgrounds in physics and mathematics could be enticed into addressing meteorological problems. He viewed positively the work of Lewis Fry Richardson (on numerical weather prediction) and Vilhelm Bjerknes of the Bergen School as steps in the right direction, but much more work was needed. It is not clear, however, just how much Woolard understood of the Richardson work. In the discussion that followed the presentation of his paper, Woolard was asked how far in advance Richardson's method could be used to forecast the weather. His response of "six to twelve hours" when Richardson himself figured it would take 64,000 human "computers" working full-time just to keep up with the weather as it happened shows a lack of comprehension of the magnitude of the obstacles that faced numerical weather prediction.[48]

While the Weather Bureau was concerned with practical, day-to-day forecasting—which had changed little since the beginning of the century (see
figure 2.1)—its meteorologists were aware that predictions would not improve
without theoretical research, and that the practical problems would not see
a solution without research on theoretical issues. Chief Charles Marvin presented a list of current research topics in meteorology to the Meteorology
Section of the American Geophysical Union (itself just founded in 1919) at
its 1923 meeting. The first item on his list: solar radiation and its influences
on terrestrial weather. As was discussed in the previous chapter, there was
great interest at this time in how variations in solar radiation controlled the
weather—an indication of the extent to which weather forecasting was seen
as tied to astronomy. Several research projects to measure incoming radiation
under a variety of conditions and in a number of different locales were being
actively pursued by Weather Bureau and academic meteorologists, along with
astronomers at the Smithsonian Institution. Knowledge of general atmospheric circulation was still sketchy. Meteorologists needed to determine how
air was exchanged across the equator and within the hemisphere. As of 1923,

Figure 2.1
Weather Bureau Forecast Office, 1926. (NOAA National Weather Service Collection,
courtesy of NOAA Central Library)

the northern hemisphere map, which had been available before World War I as a result of data sharing, could still not be produced because some observation networks in Europe had not been restored. Other meteorological fields of inquiry listed for consideration included the causes and/or events that led to the development of cyclones and anticyclones, the origin of West Indian hurricanes, and the difficulties of providing forecasts for marine and aeronautical interests. Long-range forecasting based on so-called sequences of weather conditions and periodicities of weather and climate also made Marvin's list. In the discussion that followed, his fellow meteorologists expressed their opinion that physics needed to be applied to these questions and that the discipline needed to get away from describing distributions of temperatures and precipitation without looking for explanations for their occurrence. Determining how air circulated in the atmosphere was imperative.[49]

Two years later, in 1925, Edgar Woolard again made a case for theoretical work. This time he presented the "origin, nature, structure, and maintenance of ordinary cyclones and anticyclones" as being a major unsolved problem of both theoretical and practical importance. If meteorologists were unable to determine how cyclones and anticyclones developed and dissipated, then forecasting was not going to become more accurate.[50] He realized that mechanics and thermodynamics comprised the fundamental underpinnings of atmospheric circulation. Given a complete three-dimensional set of observations, how would one use the laws of physics to determine what the atmospheric conditions would be some time in the future? Woolard did not think that meteorology could be a credible science until researchers were on the path to an "exact solution." However, neither pure mathematics nor mathematical physics nor observational meteorology had been sufficiently developed to provide an exact solution. Weather services had to settle for "inexact and fallible" empirical methods of forecasting, and theorists had to settle for qualitative or statistical explanations instead of a complete mathematical one. Bergen School meteorologists were using graphical methods to solve differential equations involved in these mathematical descriptions of the atmosphere because a direct solution was not possible. But it would be this mathematical approach that would ultimately allow insight into the mechanisms that controlled atmospheric processes.[51]

While some meteorologists were concerning themselves with atmospheric circulation, in the late 1920s others were increasingly fascinated by the possibility of long-range weather forecasting. Lines of research for forecasts that would come several days or more in advance of the event included looking for patterns in collected empirical data; determining the primary causes of unseasonal changes; examining the influence of topography that might sustain or ramify the initial changes in a weather pattern; or, when all else failed, looking at a combination of all these things. The empirical route could lead to some results in the short run, but it would take years of research before such forecasts could be made because of understanding atmospheric processes.[52] Research on the variability of the solar constant, primarily being pursued by astronomers, was also related to long-range forecasting and the possibility of climatic influence over time.[53] A difficulty hampering most long-range forecasting efforts was that the meteorology community did not grant long-range prediction any scientific standing. Consequently, students who might have been interested in pursuing this forecasting topic as a research opportunity were discouraged from entering it. Moreover, the federal government—the primary source of research funds—was not eager to provide the financial support needed to carry out such a program.[54]

The lack of credibility and funding evaporated with the passage of the Bankhead-Jones Act of 1935. An outgrowth of Dust Bowl conditions on the Great Plains, the Bankhead-Jones Act provided funds for basic research that would lead to solutions for agricultural problems.[55] As related to meteorology, it provided funds for studying long-range weather prediction and the effects of weather conditions on crops and livestock. The research attacked the theoretical basis of atmospheric circulation by examining the day-to-day change in major features and how the resulting patterns affected weather in different parts of the country. Researchers also looked for empirical clues that could be used to anticipate the future state of the general circulation pattern, i.e., the location of semi-permanent high-pressure and low-pressure areas and the wind fields that accompanied them.[56] Of course, there were differences of opinion as to what influenced the global circulation. Astronomers pinned their hopes on solar influence, including changes in sunspot patterns and other solar radiation changes. Oceanographers claimed the oceans were important to any studies of atmospheric circulation and that research into the interaction between the ocean and atmosphere was a necessity.[57] All seemed to agree that while statistics could give tantalizing hints of connections between patterns and weather phenomena, there could be no substitute for physical understanding.[58] Rossby set out a research plan for long-range forecasting during a conference in mid 1937. It included investigations of anticyclogenesis (the development of high-pressure areas), particularly as related to the influence of lateral mixing; a study of how systems become dynamically unstable; and the development of a theory of the flow patterns in the atmosphere as shown on isentropic charts.[59]

Of course, not all research paths were theory-driven. The impact of weather elements and conditions on aviation safety prompted a number of fruitful efforts. Among them was an examination of aircraft icing—in particular, research into what conditions seemed to favor its development and how it could be avoided—by the meteorologists of the Blue Hill Observatory. Since efforts to rid airplanes of ice had failed, meteorologists focused on guiding airplanes around clouds that could contribute to icing.[60] Similarly, fog severely affected visibility at both take-off and landing sites—the former being less important as long as there was no emergency forcing a take-off. However, fog at the landing site could prevent the plane from coming in. If an alternate landing field was not close by, that could doom an aircraft running low on fuel. Studies at Blue Hill addressed both fog development and forecasting in addition to research into fog dispersal. Alexander McAdie of Blue Hill thought that early morning ground fog would be the easiest to dissipate—by spraying it with electrified water. He envisioned ridding entire harbors of fog in this manner.[61]

In the early years of the society, most research in meteorology was government funded (limited though it was) and needed to show practical results quickly. As time passed and universities established meteorology programs—often in conjunction with aeronautical engineering programs—research agendas became more theoretical. Academics in the United States sought input from European meteorological research centers, either by bringing in visiting scholars or sending personnel to study in Europe.

The variety and extent of meteorological research projects grew dramatically in the interwar period spurred by the active influence of the American Meteorological Society's members. Before the society's creation, the only research publication venue was the Weather Bureau's *Monthly Weather Review*. Although not strictly an in-house organ, i.e., meteorologists outside the Bureau published their work in it, there was no opportunity for the larger meteorological community to influence the research agenda. This changed when the AMS organized and produced its own publication: the *Bulletin of the American Meteorological Society*. A mixture of short research reports, book reviews, weather-related reprints from the popular press, and gossipy news about members, *BAMS* guided the research agenda by keeping members informed. This widely disseminated meteorology community news built and strengthened professional contacts between Weather Bureau, military, and academic meteorologists—contacts that would be pivotal to the eventual success of numerical weather prediction.

Meteorology on the Eve of War

By 1938, the American meteorology community had grown larger and more cohesive as a result of the demands of World War I and the growth of the aeronautics industry in the years that followed. The Weather Bureau still employed the largest block of meteorologists—an underpaid, underfunded, and underappreciated group of people who remained entrenched in the old ways of doing meteorology. This criticism missed the mark in one crucial respect: with funding stagnant or falling (or even with funding rising but failing to keep pace with rising costs), the Weather Bureau had little choice but to stay with the old ways. The new Bergen School methods demanded increased upper-air observations, more surface observations (both spatially and temporally), and advanced training for the staff. Lack of funds alone was enough to prevent the implementation of new techniques. Apparently unable to successfully lobby Congress for sufficient funds, bureau meteorologists gamely continued to perform research when they could, with the limited funding and facilities at their disposal. They recognized that, although practical results in forecasting were desirable, basic theoretical work needed to come first before physical knowledge of the atmosphere would explain observed weather phenomena.

Starting in the 1920s, military meteorology—as represented by the Navy Aerological Service and the Army Signal Corps Meteorological Service—had been in a period of retrenchment following the draw-down after the war. Nevertheless, aviation missions had not stopped at the end of the war, and demands for services had increased even as trained personnel decreased. In 1928, the Navy's solution—to form its own graduate program at any civilian school willing to do so—had helped to encourage graduate education in meteorology. The pursuit of Bergen School methods by the Navy's aerological community leader, Francis Reichelderfer, meant that the Navy had been the first to put the air mass theory to the test in making its forecasts. However, research in the military services was generally tied to instrumentation (such as the meteorograph; see figure 2.2), and not to theory.

Despite its growth since the end of World War I, the academic meteorology community in the United States remained very, very small—less than few dozen people. Professional meteorologists numbered a few hundred—about the same as astronomers.[62] In contrast, in 1932 there were 2,500 physicists.[63] Few faculty members teaching meteorology had studied it as graduate students. Meteorology tended to be a subject tossed into geography courses or occasionally added to a physics course. It did not exist as its own academic discipline. However, the need was there and more men entered the field. Teacher training institutions, although providing non-technical instruction, added a considerable number of meteorology courses to their curricula, which helped to stir interest among all segments of the school-aged population. The establishment of the first graduate meteorology programs, first at

Figure 2.2
US Navy biplane carrying meteorograph, 13 December 1934. (NOAA National Weather Service Collection, courtesy of NOAA Central Library)

MIT in 1928 and later at NYU and Caltech, gave a significant boost to the numbers of mathematically and physically savvy meteorologists who were ready to move into research positions and advance the development of the discipline. As part of the rise of the earth sciences in the early twentieth century, the founding of the American Meteorological Society in 1919 provided the pathway for improved communication of ideas among the generally isolated individual practitioners of the atmospheric sciences.

The interwar period thus nurtured a slow, steady advancement of scientific theory, practice, and education in meteorology. But the dramatic events of World War II were soon to place great demands on meteorologists, as national defense needs stretched their capacity to respond to increasingly sophisticated operational requirements. The modest gains of the 1920s and the 1930s would soon be put to the test.

3 :: An Expanding Atmosphere: The War Years (1939–1945)

Throughout the interwar period, most meteorological training in the United States had been conducted "on the job" by the nation's three weather services. Civilians had enrolled in graduate meteorology programs starting in the early 1930s, but enrollment (and career opportunities) remained minimal despite expanding meteorological support for aviation. With the coming of World War II, the numbers of meteorologically qualified persons were insufficient to meet either domestic or military needs.

Under Rossby's direction, the University Meteorological Committee (UMC) established and coordinated an accelerated meteorology program to meet the needs of both civilian and military agencies. Military requirements led to a flood of new students, most of whom would never have considered meteorology as an academic or career field before the war, into a very small scientific discipline. The training of thousands of new meteorologists within a five-year period was an extraordinary event in the history of science in the United States that would dramatically change the meteorology community. The coordination undertaken to provide this training and assimilate these new meteorologists into the scientific community in the postwar years would prove crucial to the professionalization and advancement of the atmospheric sciences.

Changing Leadership and Expanding Instruction

As 1938 drew to a close, the Weather Bureau was moving slowly into the research arena by establishing a small unit to direct and supervise research projects. The unit's goal, according to bureau chief Willis R. Gregg, was to foster cooperation with organizations conducting meteorological research and to coordinate its own efforts with those of other institutions. Gregg was unable to deliver that message to the AMS annual meeting; on 14 September 1938 he died, at age 58, from complications of a blood clot,[1] having led the Weather Bureau for less than 5 years.

Robert Millikan, who had selected Gregg back in 1933, sprang into action when the telegram arrived advising him of Gregg's death. Within 24 hours of Gregg's passing, Millikan recommended a new chief: Navy Commander Francis Reichelderfer. The 43-year-old Reichelderfer, who had led the Navy's aerological service since the 1920s, was highly thought of in scientific circles. Millikan was convinced that the National Academy of Sciences needed to make a recommendation quickly to "prevent political influences from getting into this appointment. It is a very vital one for the scientific interests of the country."[2] A day later, Secretary of Agriculture Henry Wallace called Compton, a member of the Science Advisory Committee on the Weather Bureau, to Washington for consultations on Gregg's replacement.[3] By 5 October 1938, Millikan reported to the other advisory committee members that only two names—Reichelderfer and Rossby—had been submitted by more than one person. Other recommendations included current and former bureau employees who were in their sixties. Millikan clearly did not want an old chief—he wanted someone who was young enough to vigorously transform the Weather Bureau into an organization that could provide "effective and progressive" service.[4] Acting on a request from Millikan, Assistant Secretary of the Navy Charles Edison arranged for Reichelderfer to fly to Washington from the West Coast for an interview with the entire advisory committee. Rossby—the other contender—would also be in Washington for an interview.[5]

Compton and Chief of Naval Operations Admiral William D. Leahy, writing in favor of Rossby and Reichelderfer respectively, were glowing in their praise. Rossby, Compton wrote to Wallace, was the "unquestioned leader" in meteorology in the United States, both as an "investigator and a teacher." Indeed, Compton continued, "the majority of the present trained meteorological personnel in this country are his pupils." Further, the leader of the British delegation of the International Congress of Applied Mechanics had recently stated that "[Rossby] has now become the leading meteorologist in the world." He would, in short, be a very big asset to the Weather Bureau. Similarly, Admiral Leahy wrote highly of Reichelderfer. Although the Navy would be sorry to lose such a valuable officer, taking the long view, Leahy thought it was in everyone's best interests to have strong, positive leadership at the bureau. He recommended Reichelderfer without reservation.[6]

Millikan's committee chose Reichelderfer to aggressively carry forward the changes in Weather Bureau structure and culture initiated by Gregg.[7] Reichelderfer was then serving as the Executive Officer (the number two position) aboard the battleship USS *Utah* (BB-31) when Wallace tapped him to lead the bureau for a three year period—the equivalent of a Navy "shore tour," since Reichelderfer would be "on loan." Reichelderfer was perfectly happy in the Navy. The three-year leave had the potential to jeopardize his excellent career prospects without providing any compensating financial rewards. Despite the possible negative consequences, he accepted Wallace's offer and became the chief at the end of 1938. For Reichelderfer, the opportunity to contribute to a field "ripe for progress" was too good to turn down.[8] He in turn convinced Rossby, his long-time colleague, to take a leave of absence from MIT to become Assistant Chief of Research and Education.[9] With the appointments of Reichelderfer and Rossby arranged, the report of the Subcommittee on the Weather Bureau stated, "the direction of the Weather Bureau now possesses a prestige such as it has never before enjoyed."[10]

Although Rossby would have brought a different sort of spark to the Weather Bureau's top post, he did not want the job. Being the chief would have interfered with his own research program.[11] Rossby's mission during his three-year appointment with the bureau would be to expand its research and instructional programs. Because of the lack of educational opportunities in meteorology, the bureau had become saddled with many poorly trained people. Rossby intended to substantially raise the staff's professional standing.[12] Reichelderfer and Rossby would ensure that Weather Bureau staff members would be offered significantly expanded instructional opportunities under their leadership. As they had since World War I, aviation requirements would be the primary spur for these efforts. Indeed, in 1939, the money allotted for so-called general weather services had *declined* by $200,000 over a ten-year period, while funding for aviation weather services had continued its steady climb.[13] This funding situation, likely aided by a general fascination with all things related to flight, also occurred at a time of increasing demand by the less sexy bureau-supported sectors: agriculture, forestry, transportation, and industry.

The expansion of instructional programs had been directed by section 803 of the Civil Aeronautics Act of 1938, which required the bureau to annually send not more than ten of its members for graduate meteorology training at government expense. The bureau could send staff members to civilian or government institutions, and it planned to select and send four employees during the 1938–39 academic year.[14] However, the number of institutions offering such training was still limited. MIT and NYU offered theoretical meteorology programs at the graduate level—Rossby sent some of the Weather Bureau personnel there—while Caltech offered a master's program that was designed to meet industry needs.

Caltech's program—started in 1934 by geophysicist and seismologist Beno Gutenberg—had begun its life as a single course in atmospheric structure. Meteorology, as a branch of earth physics, was placed within the geology department. After the 1933 crash of the Navy's airship USS *Akron* and the subsequent realization of the importance of meteorology to flight, meteorology had moved under the aegis of the aeronautics department. Caltech had offered its first regular meteorology courses to seven students during the 1933–34 academic year. Enrollment had increased thereafter as airlines hired the graduates of its industrially focused curriculum. Department chairman Irving Krick did not run a theoretical department. Therefore, its offerings would prove to be of little use to the Weather Bureau. In fact, Reichelderfer detested Krick, a smug, supremely confident self-promoter who routinely argued that he could forecast weather for the entire country better than the bureau—which did not endear him to Weather Bureau meteorologists. However, a few years later, Krick's claims of long-range forecasting ability would catch the attention of the father of the US Air Force, Army Air Force General Henry H. "Hap" Arnold. Krick created a special curriculum for the nascent Air Weather Service and Caltech became its graduate program of choice.[15]

The entrepreneurial Rossby was always on the lookout for opportunities to promote meteorological instruction and establish meteorology programs. Such an occasion presented itself when Jacob Bjerknes, son of Bergen School founder Vilhelm Bjerknes, found himself trapped in the United States after Germany occupied Norway in 1939. Rossby quickly took action. He was eager to establish a theoretical meteorology program on the West Coast that would serve as an alternative to Caltech. Rossby persuaded Bjerknes to go to UCLA, which Rossby then convinced to start a meteorology program within its physics department. In 1940, Bjerknes became the chairman and Jörgen Holmboe, another Bergen School-trained Scandinavian transplant, taught dynamic meteorology. Joseph Kaplan taught his research specialty: upper atmospheric physics.[16] They were assisted by several operational meteorologists from the bureau's Los Angeles district office. The "Announcements" section of the AMS *Bulletin* proclaimed that Los Angeles had become "a leading center of meteorological professional activity," due to the additions of the UCLA program and the new district forecasting center at Caltech.[17]

While Rossby was arranging employment for Jacob Bjerknes, he was also expanding the Weather Bureau's in-house training program from the Washington central office to five district offices: Chicago, Washington, New Orleans, Denver, and San Francisco. MIT graduate and Rossby protégé Horace Byers, who was desperate to escape the Washington-area bureaucracy, offered to move to Chicago, as did Victor Starr (another MIT graduate). As Byers later recalled, he and Starr were getting tired of training bureau personnel. Byers decided to ascertain whether the University of Chicago's physics department might be interested in meteorology. He discovered that the department head had been part of the Signal Corps balloon project during

World War I, and thus would likely have meteorological interests. Soon thereafter, the physics department asked Byers and Starr to give lectures, and Chicago's vice president invited Byers for a lunchtime conversation about the possibility of establishing a meteorology program.[18]

Although Chicago officials suggested that Jacob Bjerknes or another member of the Bergen School would be a good choice to start and lead the program, Byers successfully argued for Rossby (who, of course, had trained at the Bergen School). In fall 1940, the University of Chicago program started with thirteen courses in eight subject areas. Starr, Weather Bureau meteorologist Harry Wexler, and ozone expert Oliver Wulf filled out the new team.[19] They were assisted by physicist and future Ford Motor Company vice-president of research, Michael Ference, who specialized in the upper atmosphere, and geographer H. M. Leppard. Rossby came on board as a visiting professor in the second quarter and formally left the Weather Bureau in 1941 at the end of his three-year contract.[20] With that, a "seat of meteorological education" came into being at the University of Chicago.[21]

UCLA, the University of Chicago, Caltech, MIT, and NYU thus became the centers for professional meteorological education in the United States: the "Big Five." With the exception of Caltech, the Big Five programs were dominated by Bergen School polar-front theory reinforced by the presence of Bergen-trained Scandinavians. All five schools would prove crucial to the provision of meteorological instruction in support of national defense as the United States moved closer to war.

War and Weather

At the end of World War I, military planners had assumed that the demobilized weather services could quickly come up to speed if confronted by a national emergency beyond the Weather Bureau's geographic range. In the absence of such a threat, all three weather services exchanged data and reports to increase efficiency and avoid duplication of effort.[22] As discussed above, despite new educational venues, few military meteorologists were on active duty in the late 1930s and few could imagine how important meteorology would be in executing the next war.[23]

As it became clearer that the United States would be drawn into the European war, meteorologists in all of the weather service realized that there were not going to be enough weather forecasters to meet civilian or military requirements. The extent of their potential training mission became starkly apparent after President Roosevelt's May 1940 announcement that 50,000 aircraft would be added to the military arsenal. Starting in June 1940,

Norwegian-born and -trained Sverre Petterssen conducted the first accelerated meteorology training course at MIT. The Army and Weather Bureau-Civil Aeronautics Administration (CAA) members composed this class that graduated in September 1940. Within the year, intensive nine-month courses under the auspices of the University Meteorological Committee (UMC) were either underway or planned for the Big Five meteorology programs.[24]

Although meteorology program entry requirements varied slightly by military branch, all applicants needed to be college seniors or graduates with majors or degrees in science or engineering. They were expected to have completed mathematics through integral calculus and a year of physics. Potential trainees applied to participating universities, which provided them with appropriate military forms for officer programs.[25] By late 1942, the Big Five still had insufficient applicants: 3,000 men needed to be recruited to attend the graduate level "A" courses beginning in September 1943 and March 1944. Becoming more aggressive in their search for potential trainees, the UMC's recruiting board asked universities to provide it with the names and addresses of possible candidates so they could be contacted and asked to apply.[26] The few women attendees were usually slated to backfill positions at the Weather Bureau, whose own ranks had dwindled as its forecasters were called to active duty.[27] The Navy brought women into the reserves to provide meteorological support, but unlike their male counterparts they were often required to have at least a master's degree in a scientific area to merit selection to the meteorology training program.

Military services were still facing a forecaster shortfall despite the continuing influx of young men into the training pool. By fall 1942, the UMC was eyeing another group of potential recruits: men attending junior colleges and those just graduating from high school who had strong mathematics and science skills. The junior college students were recruited through academic departments, while the high school graduates were identified at local recruiting offices or during basic training.[28] The former, needing at least one year each of college physics and mathematics, were placed into "B" courses (Pre-meteorology): accelerated six-month preparatory training to prepare them for the "A" course. The "B" course did not involve any meteorology—just prerequisite physics and calculus.[29] The high school graduates were placed in "C" courses (Basic Pre-meteorology): 12 months of mathematics and physics, plus writing and other humanities-type courses.[30] Those who performed well advanced through the other courses. Some of those selected for the "B" and "C" courses were military enlistees who had passed written tests for selection.[31]

Course directors had difficulty finding qualified instructors. The lack of qualified instructors was a major challenge facing course directors. For example, University of Washington President Lee Paul Sieg had to find several mathematics and physics instructors in order to offer the "B" course. They were not available. As it was, the university could barely cover courses already on the books.[32] Rossby recognized and acknowledged this lack of professional teachers. Even he was being forced to staff his instructor pool with recent graduates from his own program. As a result, there was a "notable lack of maturity" among those working at the University of Chicago's Institute of Meteorology.[33] Most of the available meteorologists had just completed their own graduate educations and were very inexperienced.[34]

When the United States entered the war, in December 1941, the Army Air Force had 400 weather officers and 2,000 enlisted weathermen. The Navy had 90 aerologists (weather officers) and 600 aerographer's mates (enlisted personnel). By early 1945, the AAF numbers had grown to 4,500 officers (including the two cadets being trained in figure 3.1) and almost 15,000 enlisted, while the Navy had about 1,300 aerologists and 5,000 aerographers.[35] In all,

Figure 3.1
Army Air Force Lieutenant Fred Decker (center) trains two meteorology cadets at New York University in 1944. (courtesy of USAF Historical Research Agency, Maxwell Air Force Base, Alabama)

between 7,000 and 10,000 men and women were trained as professional meteorologists and another 20,000 as observers and meteorological technicians during World War II. (More than 2,000 of the officer trainees were diverted to flight controller and navigator training, and never served as weather officers.[36]) More people received meteorological training during fiscal year 1942 than in the previous 10 years combined—and that was before the largest training classes met.[37] Even if most of these trainees returned to their original occupations, or switched into different ones, by the end of the war there would still be a marked increase in the number of professional meteorologists. By one estimate, the number of professional meteorologists at war's end was approximately 20 times greater than before 1940.[38] They would come to make substantial disciplinary changes—changes required for the development of numerical weather prediction.

The University Meteorological Committee

In January 1941, the three weather services combined their resources to begin making domestic and military meteorological support plans in anticipation of a formal declaration of war. This Interdepartmental Committee on Meteorological Defense Plans would undergo two more name changes before becoming the Joint Meteorological Committee (JMC)—the Joint Chiefs of Staff's official advisory group on weather matters—in December 1941. The JMC addressed the training and employment of civilian and military meteorologists, and also worked on standardizing weather codes to promote more efficient data sharing. They did this by arranging research in long-range weather forecasting, developing background material in support of amphibious invasions, and creating a historical northern hemisphere weather map series to be used in weather typing.[39]

Rossby had convinced military planners early on that the training project should rest with university meteorology departments, not with military and Weather Bureau in-house training courses.[40] Consequently, by fall 1942 virtually all non-governmental meteorologists were involved in the training programs to the exclusion of most other work. Unfortunately, that meant that there was no apparent effort to use scientific knowledge of weather and climate in military planning and operations. In a telegram to his long-time acquaintance Vice President Henry Wallace, Rossby offered his assistance and that of his academic colleagues. He wanted to help overcome what he saw as duplication of effort between the military services, and to develop some kind of cooperative, coordinated plan of attack for weather services to the nation. Rossby noted that he had already offered his services directly to Colonel Donald Zimmerman, who was in charge of the Army Air Force's weather services.[41] Within a couple of weeks, the War Department requested the formation of a standing committee to coordinate the recruitment and training of meteorologists for the Army and Navy air forces as well as the Army ground forces.[42]

As noted above, the University Meteorological Committee (UMC), chaired by Rossby, drew its members from the Big Five. UCLA's upper-air specialist Joseph Kaplan served as the personnel director and was responsible for recruiting. Physical meteorologist Henry G. Houghton represented MIT. Athelstan F. Spilhaus came from NYU. Paul E. Ruch of Caltech, who held an MS degree from that institution and was an associate professor, directed its meteorology program during World War II while Krick was in uniform providing forecasting services for the Army Air Force. Although the UMC was originally formed to provide guidance on recruiting and training, it would go on to influence research agendas and the professionalization of the field. By December 1942, the UMC was in full control of all meteorological training and had the full "confidence and cooperation" of the Army. The "Assistant Director of Weather" was the military head of the program, but Rossby was the organization's dominant figure.[43]

Even though meteorology training had been in progress for a couple of years, the rapid increase in student numbers dictated a more coordinated approach among participating universities. To ensure a consistent curriculum, the UMC needed to decide on course prerequisites and content. A significant debate over which mathematical approach should be taken while teaching physics ensued during a January 1943 meeting. Some faculty members recommended the simultaneous teaching of calculus and mechanics. Others thought that students should study algebra-and-trigonometry-based physics, then vector algebra, then vector calculus, and then mechanics using vector analysis. The crux of the debate focused on student preparation. While some had finished calculus, others had not. The question then became one of correlating physics instruction with the correct mathematics level so as not to lose students. Physics preparation was also weak. Despite entering the program with more than the minimum requirements, many of the men were deficient in sophomore-level college physics, which slowed their academic progress.

Civilian faculty members were also concerned about their ability to prepare students adequately if they could not control the students' waking hours. They worried that if they did not set strict limits, military authorities might appropriate students' time for military matters. Therefore, a minimum of 49 hours per week had to remain under academic control. Although the "A" course was strictly devoted to meteorology, the "B" course included mathematics, physics, mechanics, geography, and English (both written and oral). The English courses were included to prepare the men to make clear, concise radio transmissions between ground and aircraft in order to reduce misunderstandings and the accidents they might cause. All participating schools were expected to follow the assigned curriculum. If time had to be made up, it came from English and geography. Interestingly, any cuts from geography were to be first taken from climatology.[44] Although this might seem counterintuitive, physical geography would be much more important for these students than climatology. Topographic features significantly affect weather and therefore students would be well served by realizing where tall mountains, deserts, valleys, and other landforms were in relation to militarily important sites and how they could affect their ability to make a forecast. On the other hand, climatology for an area could be looked up in tables and on graphs. If time were tight in the training program, instructors assumed that students could familiarize themselves with local climatology at their ultimate duty stations.

The UMC was in place, but the Weather Directorate had been dissolved by April 1943 when Rossby expressed his concerns about meteorological support services to communications engineering pioneer Edward L. Bowles, Special Assistant to Secretary of War Henry L. Stimson. Rossby told Bowles that the demise of the Weather Directorate meant that military weather activities were rudderless once again. Furthermore, very few professional meteorologists were in positions of authority, which was not surprising given their paucity. Meteorologists had been assigned to the Weather Information Service, but it was not involved in training policies or the needs of military aviation. The Pacific campaign was being waged in tropical areas, about which little was known meteorologically and for which few training materials were available. They were desperately needed if trainees were to be competent forecasters when they arrived on station. With no War Department coordinator, specialized meteorological areas—including tropical and oceanographic meteorology (i.e., weather over oceanic areas)—could become victims of infighting between special interest groups. Additionally, the continued presence of non-meteorology-trained personnel in the decision-making pipeline was delaying prompt action on new training ideas.[45]

Rossby closed his discussion by pointing out that, as far as he knew, the United States was the only country where top-notch academic meteorologists were being used to provide basic meteorological training while no one was being tapped for policy advice. The unfortunate result: ground forces were operating without adequate weather services due to the emphasis on aviation needs. Closer cooperation between Army and Navy weather services could overcome this deficiency.[46] Although Rossby did not address it explicitly, there was another contentious point: scientists in other fields, most notably the physicists, were being used to significantly advance the war effort through their work on weapons and weapons countermeasures. They were being consulted by the highest levels of government. Because there were so many more physicists, those being employed as instructors did not constitute the majority of available scientists.[47]

For Bowles, grappling to fix defects in communications circuits handling weather data and information, Rossby's letter must have appeared as an answer to a prayer. Less than two weeks after Rossby penned his letter, Bowles appointed him as an "expert consultant" to the Secretary of War's office. Rossby's appointment letter was followed by a letter from Stimson to University of Chicago President Robert Maynard Hutchins. Stimson requested that Hutchins make Rossby available for a mission that was "vital to the war effort."[48] Bowles needed advice on how to balance the requirement for meteorological information against the ability of communications facilities to carry them in a timely manner.[49] As the war continued, Rossby's expertise would be tapped numerous times. He was asked for personnel recommendations, ideas on the best uses of newly trained meteorologists, and advice on the necessity for encrypted and coded transmission of weather data. Rossby also undertook inspection trips of the standard air routes to determine how best to support them. He coordinated meteorological support for ground forces, which had been left wanting due to the focus on aviation missions, and investigated shortcomings in instrument development and procurement.[50] Thus, Rossby came to be integrated into the highest levels of the military command structure. He tremendously influenced all aspects of meteorological services during the war. When the end of the war was in sight, he continued to use his connections to advance his personal agenda for both meteorological research and disciplinary professionalization.

Rossby attacked shortcomings in weather support being provided to military units, but many smaller, practical matters adversely affected the university training units, including difficulties obtaining current weather data via teletype because real-time data could not be sent "in the clear." For students to have data access, the universities would need to install a "secure drop" with an encrypted communications link, a guarded room, cryptographic equipment, and properly cleared personnel, because weather data were being handled as classified material. Owing to their relatively low priority, installing secure drops at the universities was not trivial. Therefore, students had to work with "canned" (i.e., old) data, which had no intelligence value and therefore could be used without security considerations. Instructors were of two minds about using canned data. Meteorology students were generally more attentive to real-time data—plotting and analyzing data from several months before was not nearly as exciting as watching the current weather unfold on the chart as it unfolded outside. Real-time data also allowed students to make the connection between what they saw on a map and what they observed outside. And yet from the instructors' point of view, old data were easier to handle. Once the instructors had analyzed a map, they knew the "right" answer and could tell at a glance where the student had gone wrong. If new data were continually clattering in on teletypes, then the instructional staff would literally be just minutes ahead of the students—not always the place a meteorology professor on a compressed schedule wanted to be when dealing with hundreds of students. Instructors had another reason for using canned data: they could select case studies instructive of different weather regimes. Live data had ties to real weather, but might not offer students the opportunity to see certain types of systems develop.

Additionally, the Army protested that it was receiving the most immature graduates because the university programs were keeping their best students on as instructors instead of sending them out to field activities. In order to protect the program's reputation, Rossby recommended that some of the best graduates be sent to field units no matter how much they were needed in the training arena. Besides being needed for on-site forecasting at bases all over the world, the newly minted meteorologists were needed to provide weather training for student pilots, the Chemical Warfare Service, and other branches of the ground forces. Rossby estimated that the chemical warfare branch alone needed about 200 weather officers. Since the War Department had made no provision for meteorological support for ground forces, Rossby had no sense of their manpower needs. Unfortunately for the meteorology program, some of the new meteorologists were being siphoned off for pilot training and never served as weather officers.

The specific needs of government units also created challenges for the meteorology training program. For instance, leaders of the Chemical Warfare Service had come to realize that they needed to determine the diffusion of smoke and fumes when either launching or receiving chemical warfare attacks. They had a two-fold requirement: assistance in interpreting research questions and help in the operations division using chemical warfare materials. However, the Chemical Warfare Service had no meteorologists on staff to provide advice. Rossby suggested that the UMC select men for weather training who already possessed degrees in chemistry and/or chemical engineering to fill this particular mission.[51] To meet the needs of its chemical warfare community, the Navy established the US Navy Chemical Warfare Training Unit at the Dugway Proving Ground (Tooele, Utah) to provide micrometeorology training focused on weather conditions within a few feet of the ground where the impact of gas warfare would be the greatest. The Army Air Force took advantage of this instruction to train their chemical officers to understand the meteorological conditions necessary for successful offensive gas operations.[52]

The UMC also spurred efforts to obtain needed information about weather conditions in militarily important operating regions. It received funds to establish an Institute for Tropical Meteorology (under the joint control of the universities of Chicago and of Puerto Rico) in Puerto Rico, which would address some of that sub-discipline's deficiencies. Additionally, several senior meteorologists were sent to Newfoundland, Greenland, Iceland, Labrador, Alaska, and India to obtain more realistic information on weather conditions for the students.[53]

The UMC's primary mission—training—continued with few hindrances. Students flowed smoothly through the courses and received operational weather station experience before being sent to their first assignment. Yet the UMC also had a research mission, and it was definitely in the applied category.

In early 1943, Army, Navy and Weather Bureau representatives advising the Joint Chiefs of Staff established meteorological research priorities. They assigned the highest priority to developing upper-level forecasting charts for aviation, and developing techniques for five-day and longer-range forecasts for strategic areas. Next in importance was the extension of forecasts for afloat and aviation missions over ocean areas by making use of observations from isolated stations. This was followed by exploiting the possibility of weather typing.[54]

The research programs being carried out by each of the Big Five meteorology departments were thus directly connected to the war effort. These efforts usually fit into one or more of several categories, including analysis/atlas, climatology, tropical meteorology, upper-air chart, and long-range forecasting projects. Analysis/atlas projects involved re-analyzing weather maps after including all available data (without the time constraints of operational meteorology). Meteorologists could then use the resulting collections of weather charts to study atmospheric patterns and the resulting weather. Climatology projects involved compiling many years of observational data from sites that were important to military operations (for example, Greenland, Iceland, and Europe), and determining long-term averages for temperature, precipitation, pressure, and other weather elements. This information was used by military planners to determine, for example, the best locations for landing strips, and the best (or worst) times for launching certain kinds of military operations.

Tropical meteorology studies were important because a considerable portion of the Pacific Theater was in the tropics, for which there was very little meteorological knowledge. Researchers gathered as much observational data as possible and analyzed the resulting patterns to determine forecasting rules. The last category—upper-air observations and their study—was important because it was not possible to determine the dynamic structure of the atmosphere by just looking at surface data. With the inclusion of large amounts of upper-air data, meteorologists could effectively study the general circulation of the atmosphere for the first time. They could also use the resulting knowledge to better predict flight conditions and to make attempts at longer-range forecasts, i.e., over several days instead of just one or two.

In short, these wartime research efforts were focused on improving the success rate of military operations by incorporating the latest meteorological techniques—including new instruments—and knowledge in geographical areas that had previously been outside routine military operating areas. Methods of providing long-range forecasts, including an analysis of ozone content and its relation to general atmospheric circulation patterns, were especially important to military planners.[55] This preliminary work driven by military requirements during the war served to advance theoretical studies in general circulation as well as long-range forecasting after the war.

In late 1943, with the war's end not yet clearly in sight, UMC members were considering its future. With sufficient meteorological manpower trained, its instructional role was ending. The time had come to turn their attention elsewhere and create a more permanent entity. Having been responsible for the professional training of the majority of active meteorologists who would be practicing the science in the postwar years, committee members wanted to ensure the appropriate employment of these new additions to the community. Rossby was convinced that the UMC should have an important role in promoting and developing the meteorological sciences.[56] In that vein, Rossby proposed that the UMC become the meteorology section of the National Defense Research Committee (NDRC). Created in 1940 in response to the national emergency and focused on weapons research, a year later the NDRC was subsumed under the Office of Scientific Research and Development (OSRD), which was created to mobilize scientific personnel for the war effort.[57] If such a meteorological section were established, it would be responsible for sponsoring meteorological research and have the funds to do so on a large scale. When Rossby approached Bowles with this idea, the latter strongly opposed it. Rossby was concerned that Bowles thought the Army had enough funds to support "all legitimate meteorological research."[58] Apparently the only research worth doing would be of direct benefit to the Army Air Force.

Rossby also suggested, based on information he had received from a number of sources, that the UMC could become the basis of a new professional society that would either supplant or augment the American Meteorological Society. As configured, the UMC only spoke for universities. The Weather Bureau only represented civilian sector government meteorology. And the AMS could not speak for meteorology as a whole because of the way it was organized—presumably because the AMS mixed amateurs with professional members.[59]

Some very important questions arose for Rossby and other leading UMC members. With the creation of a large cadre of highly educated meteorologists, who would be considered as a "meteorologist" in the postwar years? Who, or what, would control the meteorological research agenda? Who were the possible patrons? How might they influence the conduct of research? How would the burgeoning private sector meteorologists be controlled? What entity would be responsible for protecting the professional standing of the scientific community by licensing practitioners? How could meteorology be sold as a technical profession? All of these questions conflate to a single theme: the professionalization of meteorology.

Professionalization of the Meteorological Community

In the United States, meteorology had always been unique among scientific disciplines because the vast majority of its practitioners were employed not in universities or industrial settings, but by the government. Those employed by the government were almost exclusively attached to the Weather Bureau. Although there were other scientific agencies, including the Coast and Geodetic Survey, the Geological Survey, the Bureau of Mines, and the Naval Observatory, none of these organizations employed the majority of the scientists in their disciplines. However, with the atmosphere extending beyond the boundaries of any state or region, and data collection and processing being beyond the capability of a business concern, by national necessity the Weather Bureau had a stranglehold on American meteorological practice. Thus, before World War II, meteorologists in the United States were divided into two very distinct camps: the academic theoreticians, who were generally found within university physics or geography departments, and the forecasters, who worked predominantly for the Weather Bureau. While the former had advanced degrees from colleges in the United States or Europe, the latter had received most of their training "on the job." These two groups intermingled very little. The theoreticians considered themselves to be practicing a science and thought the forecasters were pursuing an art only peripherally related to science. For the most part, academics were seldom, if ever, involved with making forecasts. Despite these differences, anyone who was interested in the study of weather—for academic or practical purposes or just out of personal desire—was eligible to be a member of the American Meteorological Society. No distinction was made between those who were theoretical, applied, or amateur meteorologists. Indeed, AMS membership was almost evenly split between amateurs and those who were either theoretical or applied meteorologists. With membership in 1940 at a little more than 1,400, thousands of potential new members with formal meteorology training stood ready to change this mix.[60]

The academics composing the UMC considered themselves and their colleagues to be professionals. "This Committee," Horace Byers wryly observed, "may consider itself as perhaps more deeply engaged in some of the better aspects of meteorology. . . . Certainly meteorology in this country outside of this esoteric bunch is a small proposition."[61] Besides the academics, they acknowledged that there were just a handful of professionals scattered within the ranks of the Weather Bureau—at most perhaps 200–300 (or less than 10 percent of the total). Byers estimated that only 25–30 percent of those holding professional grade appointments at the Weather Bureau possessed any kind of college degree. Indeed, the bureau preferred to train its own forecasters. Prospective forecasters needed to have completed 2 years of college with mathematics and physics and have passed a placement test. Alternatively, they could have one year of college and an outstanding record at the bureau—usually as sub-professional plotters or observers. Thus, by the academics' definition the majority were not professionals.[62] In contrast, everyone they had trained during the war was a professional by virtue of course work no matter the extent of their practical experience. Uncomfortable with requiring new meteorologists to meet a higher standard of professionalism than current practitioners once the war was over, in 1943 the UMC members pondered who would have the authority to make that kind of decision.

The AMS was open to anyone who paid the annual dues of $3.50.[63] The UMC members anticipated that with the end of the war, the combined effects of a rapid increase in commercial aviation and the UMC-established meteorology training centers would result in a tremendous growth of interest in meteorology. People interested in meteorology would want an organization for amateurs. The AMS would be a good place for them—the "National Geographic Society of meteorology," as Rossby dubbed it.[64]

The professionals, no matter for whom they worked, would need their own separate organization. The UMC members wanted a professional meteorological society (much like the American Society of Mechanical Engineers) that would guarantee standards: standards for entering the profession, and standards for remaining within it. By expanding their group to include other academics and those with advanced degrees in meteorology, they could create a professional society that would be responsible for setting educational standards, accrediting university curricula, and licensing private consultants. It had to be a strong society whose words carried weight so as to "do away with the embarrassment already existing" in the profession—an embarrassment stemming from the perception of both the wider scientific community and populace at large that meteorology was not a scientific endeavor. Unfortunately, such a society would exclude a large number of current practitioners. The idea of setting up a competing group struck Byers as "snobbery."[65]

The point, as Rossby saw it, was one of rampant professional opportunism. He was convinced that when the war ended and the men returned home, meteorology would "blossom out as a field of consulting meteorological engineers." Without adequate professional standards, Rossby worried, a "lot of people" who did not possess minimal professional educations would set themselves up as consulting meteorologists.[66] A licensing venue had to be established to prevent that from happening. The Weather Bureau was a possibility; however, the UMC members did not think the bureau was capable of maintaining professional standards within its own ranks, much less among a larger group of meteorologists. Furthermore, Rossby did not think that Reichelderfer (who had stayed on as bureau chief long after the promised three-year appointment) wanted to step into the licensing void.[67] If the UMC members established an alternative professional organization, how could they license meteorologists and not extend those same licenses to Weather Bureau members who were already serving in professional positions? How could they say "no" to private sector meteorologists and not say the same thing to Weather Bureau personnel? After all, there were already many people working as "meteorologists" in private industry who only possessed high school educations.[68] It appeared that setting up a separate professional society was the only method of controlling the potentially embarrassing position of non-professionals providing meteorological services to an unsuspecting public.

But Rossby had a conflict of interest. He had been nominated as the next AMS president. (And was later elected.) As AMS president, Rossby could hardly agitate for a new professional society. That meant that AMS leaders would need to modernize its existing structure to transform it into a more professional society.[69] By January 1944, Rossby concluded that there was no need for another organization whose only purpose would be to safeguard professional standards, ethics, and privileges—much like a trade union. The AMS Council met four times on this subject between late January and mid July 1944. The councilors decided to promote the public acceptance of meteorology as a technical profession by establishing standards, issuing a new technical journal, bringing meteorology to the attention of industry, and providing a placement service for meteorologists. They would create a new category of professional members who would be "actively engaged in professional phases of meteorology who see their obligation to the science and who are therefore willing to support measures that will apparently best meet these responsibilities and insure to the general benefit."[70] By late 1944, Rossby joined Henry G. Houghton in leading the AMS to become a more professional scientific society. As part of that process, the AMS founded the *Journal of Meteorology* as the technical counterpart to the *Bulletin of the American Meteorological Society*.[71]

Another matter related to the professionalization of the field concerned university curricula. UMC members feared "wild growth" in meteorological institutions before they could get their professional organization started. However, it would be difficult for the new organization to "meddle" in the business of professional meteorology schools if they could not decide what constituted an acceptable course of instruction. Rossby proposed that representatives from the Weather Bureau, the Army, the Navy, and UMC draft a statement outlining minimum content and staffing requirements for a legitimate meteorology curriculum. The UMC would mail the statement to all college presidents, "many of whom are now thinking of establishing a professional course in meteorology."[72] These expectations were clearly emerging from unrealistic, and wishful, thinking. Certainly there were universities planning to establish meteorology departments at the end of the war. Among them were the University of Washington and the University of New Mexico. The former already had an oceanography department and wanted to expand into meteorology as a related area. Indeed, University of Washington faculty member Philemon E. Church, working with Rossby in Chicago, had advised President Paul Sieg that the university needed to make its move into meteorology if it wanted to secure the possibility of expanding into the field after the war. That was the primary reason Sieg had actively pursued offering the wartime "B" course.[73] Even if Rossby and the UMC were anticipating unfettered growth of meteorology programs, at least the University of Washington realized that demand would be limited.

Meteorology at War's End

By mid 1945, the meteorology community in the United States had been transformed. The military exploitation of aviation assets to transport armaments, personnel, and material had given an unprecedented stimulus to the science of meteorology. To ensure the safety of aviators and their aircraft from weather vagaries, meteorologists had sought and obtained the establishment of worldwide reporting and forecasting stations, an observational network that would prove crucial to postwar disciplinary development. Meteorological sectors directly related to the war effort—tropical, high-latitude, oceanographic, and high-altitude—had received a significant boost due to government-financed research. And, of course, large numbers of newly trained meteorologists had entered the field.[74]

For the United States, this was an especially profound change. Before the war, Germany and Scandinavia had considered meteorology and other geophysical sciences to be more important than had the United States.[75] Indeed, most research advances in meteorology had come at the hands of Scandinavians working under the inspiration of Vilhelm Bjerknes and the Bergen School. Although some ground had been gained during the interwar years in the United States, the Great Depression had not been a good time to advance new academic fields.[76] Progress that was made was very much influenced by the Scandinavians—at first just Rossby, later Jacob Bjerknes, Sverre Petterssen, Jörgen Holmboe, and Bernhard Haurwitz—all Scandinavians and Germans caught in the United States when the war broke out. Their leadership was crucial not only in the training of thousands of military men and women, but also in attracting those who might have gone into the physical sciences or engineering had the war not changed their plans. As a consequence, not only did the professional meteorological community grow from a total of 400 persons before the war to 6,000 afterwards—a 1,500 percent increase—but those new members tended to come from physics and mathematics backgrounds that led them to take a very different approach to the science of meteorology.[77] These were men who depended on physical laws and mathematical manipulation to define the state of the atmosphere, rather than men with a sense of the atmosphere based on gut instinct. They were looking for mathematical rigor. If they did not find it, they expected to create it.

Equally important was the new perception of meteorology among the scientific community at large and the general public. There was no mistaking the importance of meteorology to the war effort. Military operations—airborne, amphibious, ground, and afloat—all depended on accurate weather forecasts.

Not only did airborne operations require weather forecasts for safety of flight; the pilots and their crews needed to know in advance if they could count on clouds for cover, or if those same clouds would prevent them from finding their target. Likewise, amphibious missions depended on accurate wave and surf forecasts for safe beach landings and successful outcomes.

One particularly well-known instance of the importance of weather forecasting was during the planning of Operation Overlord (the invasion of Normandy on 6 June 1944). A multi-national team of meteorologists stationed in Britain coordinated weather as well as sea, swell, and surf forecasts several days in advance—an extremely difficult prospect considering the available data and techniques of the time. The conditions under which military operations had been carried out were made known to the general public through newsreels and radio and newspaper reports. No longer as quick to make snide comments about the local "weather guesser," citizens became more accepting of the discipline as a scientific profession and the meteorologist as a "reliable professional man."[78]

Community leaders realized that they had to capitalize on this newfound respect for their science and do so quickly. Under Rossby's leadership, the AMS acted swiftly to take its place beside other engineering and scientific professional societies. To guarantee that meteorology did not lose wartime gains and that it continued to advance, the AMS sought broad exposure of the discipline's possibilities. The value of meteorology to aviation and agriculture was already well known. Rossby wanted many more industries to recognize that meteorologists could make their businesses more profitable. To make this happen, the AMS took two important steps. First, it became a potent factor in soliciting funds and fellowships for research. In this way, the science could—as Sverre Petterssen put it—be "lifted out of a state of neglect and place[d] on a level of prominence amongst the other physical sciences."[79] Second, it established strict ethical standards to prevent the exorbitant claims by those who might wish to profit at the expense of an uninformed public, thus leaving the meteorological community open to criticism that could destroy the very gains made toward credibility during the war years.[80] In particular, the society was worried about private sector meteorologists who sold very-long-range forecasts based on dubious scientific reasoning to agricultural, utility, and other industrial interests who made business decisions based on these predictions.

"There has been little except war and the needs of the general public to promote advancement of [meteorology]," Army Air Force Brigadier General D. N. Yates, Chief of the Air Weather Service, later noted. "There has been practically no incentive to individuals for entrance into the field of meteorology on a career basis."[81] As unfortunate as the war had been, it opened the door for huge advances in meteorology. Horace Byers may have summed it up best: "It is an unfortunate characteristic of meteorology that its great forward strides depend on disasters. Catastrophes and wars result in increased meteorological financing and activity and World War II was an outstanding example of growth bred from disaster."[82]

World War II had expanded the atmosphere for the meteorology community—more money for research that extended knowledge into new areas, newly designed equipment, more observing stations, and more professional scientists. This "critical mass" of well-trained, ambitious, and forward-looking men became meteorologists in the postwar era—a time when virtually anything seemed possible scientifically. They were ready to take the field from a small, marginalized, and sometimes scorned, scientific backwater to a discipline of importance within the sciences and the realm of public opinion. Within a few months of VJ Day, the technology that meteorologists needed to create the scientific breakthrough of the century arrived on their doorstep: the electronic digital computer.

4 :: Initial Atmospheric Conditions: Scientific Goals, Civilian Manpower, and Military Funding (1944–1948)

By 1944, the United States was within reach of military victory. The meteorologists who had been training thousands of men to support military missions were faced with empty classrooms. For them, the end of the war appeared to be in sight. Many meteorologists were more than ready to abandon the applied meteorological questions, which they had pursued to support the nation's defense, for more theoretical pursuits. Their interests and concerns were not just limited to research topics. They extended to research *funding*. Government funding had dominated the war years. Would the free flow of money continue after the war? Other scientific communities were facing the same kinds of questions. Physics and engineering disciplines in particular were benefiting from the needs of the war effort, stimulating rapid, ground-breaking advances in radar, electronics, proximity fuses, and nuclear power. Applied physics had also stimulated development in another significant area. The extensive calculations that had to be undertaken to aim very large guns and rockets (fire-control solutions), had provided the impetus to create primitive electronic computers.[1]

One of those involved in developing what became the modern computer era was the distinguished mathematics prodigy John von Neumann of the Institute for Advanced Study in Princeton. Developer of the theory of games in the 1920s before immigrating to the United States from Hungary, von Neumann had worked on the Manhattan Project, among other efforts, during the war.[2] He sought to pursue his goal of a digital electronic computer once the war was over. This new stored-program computer would allow for significantly faster solutions of complex mathematical problems—particularly those that had non-linear solutions solvable only by numerical analysis techniques. Instead of taking days, weeks, or months of work by human "computers," these new "electronic brains" could produce a solution in hours or days. Thus, investigators could rapidly revise formulas, change variables, alter input, and re-compute as many times as necessary to reach the desired solution.

Despite the atomic bomb and the subsequent concerns about the dangers of radioactivity, the prevailing view in the United States after 1945 was that the sciences had had a positive influence on the outcome of the war.[3] This reinforced the tremendous faith of Americans, already evident in the Progressive Era (circa 1890–1917), in the ability of science to solve all difficulties—natural or man-made. Therefore, it seemed very reasonable for people to be able to control nature and their environment.[4] The atmosphere and its processes constitute the ultimate environment to control. If people could outwit the weather—prevent droughts and floods, enhance rain and snow, disperse fog, reduce hail damage, dissipate or change the paths of hurricanes, and prevent tornadoes from forming—that would be a huge achievement. Never before had such a possibility seemed within reach.[5]

But before people could control the weather, they would have to thoroughly understand the atmosphere's physical processes, and they would have to be able to consistently make accurate forecasts. In 1945, theories defining the atmosphere's general circulation were very weak. Forecasting techniques remained primitive. Nevertheless, the steps taken to aid weather forecasting during the war, such as expanding the data network and adding many more upper-air observations (figure 4.1), could now be exploited for theoretical work.

Figure 4.1
Army Air Force meteorologists prepare a balloon launch in Iceland, circa 1944. (NOAA National Weather Service Collection, courtesy of NOAA Central Library)

Figure 4.2
Carl-Gustav Rossby (courtesy of MIT Museum)

The return to peacetime helped fuel a dramatic expansion in many physical science fields, including studies of the atmosphere. Meteorologists from all parts of the community—academic, Weather Bureau, and military—were free to tackle long-term projects of importance to the overall advancement of the atmospheric sciences. The Big Five meteorology departments moved ahead with expanded research agendas. Weather Bureau Chief Francis Reichelderfer anticipated taking back some of the traditionally civilian roles that had been usurped by the military during the war. Military agencies looked at ways of influencing scientific development. In the immediate postwar period, each group maneuvered to solidify its position in a strengthened professional community.

The emergence of the electronic digital computer would prove to be a vital ingredient for the meteorologists' advancement of their discipline. However, the forward-looking efforts of Reichelderfer, Carl-Gustav Rossby (figure 4.2), and the military meteorologists were what moved numerical weather prediction forward.[6]

The Postwar Research Agenda for Meteorology

Theoretical meteorologists, such as Hans Panofsky at NYU, Horace Byers at the University of Chicago, and Henry Houghton at MIT, were eager to return to their own research projects that had been put on hold since the beginning of the war. At the same time, they were concerned about who, or what, might influence or control the postwar research agenda. During wartime, the research agenda had been heavily influenced by military requirements. The academic meteorologists were now confronted with the possibility that postwar research would be controlled by the government as well: not as overtly, perhaps, as during the war, but certainly as a result of making funding available through contracts.

Although generous funding for basic research in the postwar era would indeed become available—first through the Office of Naval Research, and later through the National Science Foundation—this was obviously not known in early 1944. The academics on the University Meteorological Committee were panicked by the thought of continued government control of their research projects. Before the war, a large percentage of meteorological and other scientific research had been funded by private sources. Leading a discussion on potential research funding during a UMC executive meeting, Byers argued that funding from private sources appeared to be on the wane, with government funding taking its place.[7] If government agencies were providing the funding, they would in turn dictate the problems to be solved, present them to universities, award contracts, and expect results. He did not anticipate that funding would be awarded for general research. The path would be laid out for a specific result, and the contract awarded to the school best equipped to provide it.[8] Byers's proffered scenario worried these meteorologists who were counting on the opportunity to conduct fundamental research—they needed the freedom to explore and follow where their research took them. In general, these academics did not view decision-making personnel in the military weather services or the Weather Bureau as being cognizant of how that research was done.[9]

Although some government agencies (particularly the National Advisory Committee for Aeronautics) had allowed considerable latitude in how contracts were handled, the Army had not.[10] Jörgen Holmboe of UCLA protested that what the Weather Bureau really wanted were improvements, not basic research. Therefore, university meteorology departments could help the bureau without getting tied down in extensive research projects that would interfere with departmental interests. Nevertheless, Byers cautioned that while it might be acceptable to help the bureau and the Army with their projects, keeping academic departments fully occupied with contracted research would greatly reduce research freedom—"the life blood of any university."[11] Houghton was blunter still. Taking on government contracts, he growled, was "selling out."[12]

Despite the prevailing evidence that private funding was a thing of the past, the UMC meteorologists were not sure that government subsidies would be a sure thing after the war.[13] Beno Gutenberg of Caltech flatly rejected the idea of government funding. He argued that once the war was over, private foundations would resume their prewar roles as patrons of basic scientific research. The government would fund research by its own people in its own labs. Gutenberg was not even convinced that any government funding would be available. He thought Congress might well divert funds earmarked for scientific research to other needs. As Joseph Kaplan contended, if they could get the private funding scenario in place well in advance of the war winding down, there would be less interference in their research agendas. With research freedom preserved, they could make more progress. But Byers and others remained skeptical. The shift to government funding had preceded the war, they pointed out, and that pattern could continue.[14]

The Weather Bureau's research budget had always been small. What had been in the pot had been quickly placed under military control during the war. As in most situations involving money, once an organization has gained control of funding at the expense of another, the latter rarely recovers it. Indeed, in the mid 1940s, the bureau had research questions in need of answers, but no funds to pursue them. The military services also had questions that needed answers, but they had plenty of money to spend. To avoid being cut out of the picture entirely, Reichelderfer recognized that he would need to place himself in a position to influence the allocation of funds.

A "top-ten list" of the "most useful research to bring results in the shortest amount of time" provides a tantalizing piece of evidence that the Weather Bureau was trying to prioritize, and perhaps influence, meteorology's postwar research agenda. The list, a result of a 1944 survey of bureau, military, airline, and university meteorologists, is interesting both for what it includes and what it does not. Seven of the ten items were related in some way to forecasting (development of forecasting rules, studies of orographic influences, studies of factors controlling movement of high-pressure and low-pressure areas). Two of them dealt with obtaining better upper-air data from radiosondes. And the last one called for descriptive studies of convergence, divergence, vertical motion and vorticity, i.e., physical processes in the atmosphere.[15]

This period was ripe for theoretical development and research in meteorology. Networks for recording surface and upper-air conditions were now at higher resolution because of the war. More scientifically trained meteorologists possessed advanced technical capabilities. And yet not one theoretical topic appeared in the top ten. Granted, one might not expect theoretical projects to "bring results in the shortest period of time." Yet this seems to be another indicator of the divide that existed between the theoretical and applied sides of the meteorological house. Meteorology's scientific advancement depended on developing a mathematical and physical theory of atmospheric circulation. It did not appear on the list. Despite comments to the contrary as meteorologists later reminisced on this period, there were no projects that indicated an interest in using a numerical approach to solving the non-linear equations defining atmospheric movement. Knowledge of the difficulty in solving such equations was probably one of the reasons. But as the war ended, the means for their solution awaited.

High-Speed Computing Meets Meteorology

Numerical weather prediction depended on the availability of a high-speed electronic computer. The creation of such a machine began in Philadelphia during the war. John W. Mauchly and J. Presper Eckert, two electrical engineers at the University of Pennsylvania's Moore School, started working on the ENIAC (Electronic Numerical Integrator and Computer) in June 1943 while under contract to Army Ordnance. ENIAC was designed to compute firing tables much more quickly than was possible with calculating machines.[16] In April 1945, just a few months before ENIAC's delivery to the Army's Aberdeen Proving Ground, Mauchly visited Weather Bureau headquarters to ascertain the possible meteorological uses of high-speed sorting and computing devices. The bureau's Assistant Director for Scientific Research, statistician C. F. Sarle, made an uninspired suggestion: use the computer to sort IBM cards punched with data for climatological studies. (Owing to a shortage of personnel, the bureau was chronically behind in computing climatological data.) Sarle also expressed interest in weather map extrapolation, i.e., the creation of a new forecast map by shifting weather features in the direction of general atmospheric flow. Such an extrapolation method would do the same thing that human forecasters were already doing—moving frontal features several hundred miles downstream, depending on the velocity of winds at the atmosphere's steering level. As such, it did not incorporate the use of physical laws in anticipating atmospheric motion. Mauchly was not sure that his new machine, EDVAC (Electronic Discrete Variable Automatic Computer), could extrapolate weather maps, but he did point out that it would be able to solve partial differential equations, i.e., the type describing atmospheric motion. Sarle was not interested. The Weather Bureau—burdened by increasing demands and shrinking manpower—was primarily interested in automating data handling and statistical computations.

A much different response greeted Mauchly when he visited the Air Weather Service. There he met former, and returning, Weather Bureau meteorologist Major Harry Wexler. Recall that Wexler had been hired from Rossby's MIT program after the Science Advisory Board urged the bureau to adopt Bergen School methods in the mid 1930s. Indeed, Wexler had been one of three new PhDs familiar with air-mass analysis hired to spread the technique throughout the bureau. Thus, he was not only interested in applied forecasting—he was interested in meteorological theory. The enthusiastic Wexler recognized the importance of applying a machine such as EDVAC to the integration of the hydrodynamic equations. He introduced Mauchly to other weather officers who were working on a variety of meteorological topics of interest to the Army Air Force in the days just before Germany's fall. They, too, were convinced the machine could have a very important use in weather forecasting.[17]

The difference between the perceived uses of the computing machine by bureau and AAF personnel in spring 1945 is striking. Sarle—bogged down with data waiting to be analyzed for climatological studies and very few people to do it—saw only pedestrian uses. In stark contrast, Wexler and his mathematically savvy AAF colleagues immediately identified an application to forecasting—*the* major meteorological undertaking for both the military and the bureau. Upon returning to the Weather Bureau at the end of 1945, Wexler would vigorously pursue this new technology.

Mauchly was not alone in trying to convince the Weather Bureau of the possible uses of the computer for meteorological purposes. Vladimir K. Zworykin of the Princeton RCA Laboratory also envisioned meteorological applications. The inventor of the electronic-scanning television camera, Zworykin, a physicist, was involved in the development of meteorological instruments at RCA and had become attracted to weather problems, including, perhaps, the ultimate one: weather control. During an evening lubricated by a liberal supply of vodka, he explained to an astonished Wexler that nascent hurricanes could be snuffed out by spreading oil over the ocean's surface and setting it afire under billowing convective clouds. Zworykin thought this would "bleed" the energy out of the system and prevent hurricanes from having sufficient energy to form.[18] Despite his incredulity over Zworykin's hurricane-killing ideas, Wexler remained fascinated by Zworykin's concept of using computers to make numerical forecasts.

Reichelderfer first heard of Zworykin's proposal for the use of "modern electronic devices" in meteorological analysis during a September 1945 visit to the RCA Lab. Much interested, he requested a copy of Zworykin's forthcoming written proposal.[19] Edward U. Condon, an ambitious physicist who was Director of the National Bureau of Standards, was also curious about this potential use of computers. Having already obtained copies of the Zworykin proposal, Condon forwarded copies to Reichelderfer and suggested that they cooperate on work with electronic computers. Reichelderfer observed at the time that, although Zworykin's proposal was unproven, it "should not be taken lightly."[20] In early December, Reichelderfer pursued the possibility of using electronic computers for meteorological analysis and extended forecasting by inviting Zworykin to bureau headquarters to discuss a potential collaboration in more depth.[21] As that letter was leaving his office, Condon contacted Reichelderfer and suggested that they also invite von Neumann.[22]

Originally planned for the end of December, the conference was rescheduled for 9 January 1946. Attendees would include a small number of bureau and military meteorologists in addition to representatives of the Bureau of Standards. Reichelderfer noted that Zworykin's ideas constituted a "startling, but *noteworthy* proposal."[23] In his invitation to von Neumann, Reichelderfer wrote that the purpose of the conference was to discuss "the ways and means for improving the techniques of weather analysis and forecasting."[24]

While the meteorologists voiced tentative interest in this proposed computer's application to weather forecasting, the Navy's Office of Research and Invention (ORI) was already getting out its checkbook to fund von Neumann's computer. During their meeting with IAS Director Frank Aydelotte, the Navy's Chief of Naval Research, Admiral Harold G. Bowen, and the head of the ORI, Captain Luis de Florez, were "very enthusiastic" about the computing machine. Their "purely scientific" interest came with a commitment to make a substantial "no-strings attached" contribution to the effort.[25] Whatever de Florez had in mind when expressing enthusiasm for this plan, the Navy's claim of "purely scientific" interest is questionable. Several years later, de Florez publicly proclaimed himself a strong supporter of weather control.[26]

With a well-placed leak to the *New York Times*, the Navy revealed the possibility of a weather-predicting computer within two days of the meeting brokered by the Weather Bureau. Sources told the *Times* that participants in that meeting had discussed a new super calculator that not only would be able to predict the weather but would make it possible to "do something about the weather" by using "counter-measures" against unfavorable conditions. Navy meteorologists thought sufficient theory existed, but the complicated calculations could not be solved quickly. The new computer would reduce the time required to find solutions. The *Times* reported that some scientists thought that the threat of tornadoes, hurricanes, and other severe weather could be reduced with advance knowledge. For example, atomic energy (i.e., nuclear weapons) might be used to divert hurricanes away from populated areas.[27] The Weather Bureau was interested in analysis and forecasting applications. The Navy, which heretofore was only involved in funding the computer, now seemed to emphasize the weather-control aspects of Zworykin's proposal in its off-the-record comments.

Reichelderfer and other conference participants, thinking their meeting was confidential, were very unhappy with the coverage by the *Times*.[28] The War Department's Ordnance Research Office thought the newspaper had violated military security.[29] A puzzled Zworykin could not understand why the Navy had released the information without consulting anyone.[30]

Why the Navy leaked the story almost certainly was related to mustering support among Navy leaders for developing a meteorological application for von Neumann's computer. Navy meteorologists, like their counterparts in the Weather Bureau and in the Air Force, recognized that the computer had the potential to speed up the availability of predictive charts and to increase their accuracy. This new tool would allow on-site forecasters to spend more time on locally tailored weather prediction. For military leaders, weather was of concern only when it hampered operations. When it was not troublesome, no one gave it a second thought. To obtain continued support from war-fighting interests, meteorologists would need something more appealing than a faster forecast. Weather control, with its possible application as an offensive and defensive weapon, was clearly very appealing. Thus, the leak indicated that a comprehensive meteorological theory existed (when it most certainly did not) and emphasized the weather-control aspects of Zworykin's proposal. In order to sell a project that could forecast or control the weather, the meteorologists had to develop a plausible atmospheric theory.

Though ruffled and embarrassed by this unanticipated public-relations fiasco, Reichelderfer pressed forward to pursue the possibilities that electronic computing might offer. Wexler visited Zworykin and von Neumann in Princeton to discuss potential computer-solvable meteorological problems. Having no meteorological background, von Neumann needed advice on the mathematical and physical requirements that had to be considered.[31]

So did Mauchly and Eckert when Wexler sounded them out on the feasibility of designing an ENIAC-type machine to forecast the weather. They were convinced it could be done "once specifications were laid down by meteorologists."[32] But this was going to be difficult. Neither "electronic engineers nor the meteorologists" were able to answer the question "How can the electronic computer be applied to meteorology?"[33] Establishing the specifications would be impossible without first determining the extent of the meteorological questions. Ascertaining those questions would be difficult. The Air Weather Staff wanted to know if Wexler had any ideas, other than a reconstruction of the Richardson method, which they could think about and discuss.[34] Lewis Fry Richardson's World War I-era attempt at numerical weather prediction had been to solve the "primitive equations" of the atmosphere by making one six-hour time step and then doing all the calculations by hand (a time step being the change in time in the equations of motion over 6 hours). His results appeared in the 1922 book *Weather Prediction by Numerical Process*, wherein Richardson laid out his theoretical ideas and provided a complete example. When he was done, Richardson's forecast called for a pressure change of 145 millibars—an unrealistic value in comparison with the 1-millibar change in pressure actually observed.[35] (For comparison, a hurricane passing directly overhead may lower the pressure by 30 millibars.) Although widely reviewed, Richardson's book had attracted little attention at the time because his method was much too labor intensive. But in 1946 Richardson's approach was briefly considered as a first point of departure. For his part, von Neumann expressed his intent to examine the fundamental theories of meteorology—a necessity if the computer was to be able to solve atmospheric problems. Further, von Neumann expected to spend about 25 percent of his time on the meteorology part of the project—a figure that Rossby thought would more realistically amount to 5 percent, in view of von Neumann's other obligations.[36]

After meeting with Wexler and others in early February in Princeton, von Neumann turned to Rossby for advice. Von Neumann informed Rossby that he was "considerably attracted" by the research problem presented by the general circulation of the atmosphere, and proposed that they first try solving it in its most "simplified and schematic form." He wanted to consider a homogeneous, rotating earth that included some corrections for the amount of solar radiation received by latitude and assumed zonal symmetry, i.e., that physical data were independent of longitude. Did Rossby think that an approach using partial differential equations to describe the general atmospheric circulation would be reasonable? Because it was unlikely that he could get to Chicago in the near future, von Neumann asked Rossby to come to the IAS so they could discuss it in more detail.[37] Rossby accepted.

Rossby's discussions with von Neumann at Princeton furthered advanced the meteorology-computer connection, and he immediately reported the meeting's results to Reichelderfer. Rossby recognized that von Neumann was interested in hydrodynamic research questions and their answers, i.e., researching the general circulation of the atmosphere, but that he was not interested in the kind of empirical correlations over time and space that could have an immediate practical impact on weather forecasting. Although not aiding forecasting in the near term, the development of working models would allow meteorologists to change input variables like solar radiation to evaluate their effect. Rossby thought meteorology would be better served by letting a team work on the general circulation problem—theory development—first. Solving the equations of motion as they related to weather prediction—an application—could come later.

A master at recognizing fruitful opportunities, Rossby shrewdly viewed von Neumann's newfound interest in theoretical meteorology as a potentially huge asset to meteorological progress, and wanted to "stimulate him further" by surrounding him with a "small and versatile" group of theoretical meteorologists who would serve to provide the foundation for this new computational approach. At a minimum, Rossby wrote to Reichelderfer, they should find "some highly competent young man" to help von Neumann. However, all the young meteorologists he knew were already engaged in their own work. Instead, Rossby proposed forming a team of meteorologists that could assist in the project. His proposed list included Walter M. Elsasser, a specialist in atmospheric radiative processes with the Princeton RCA labs; Chaim L. Pekeris, a former Rossby student, working on radiation and hydrodynamics; AAF Captain Gilbert Hunt, a wartime-trained meteorologist and mathematics doctoral student at Princeton; and someone familiar with large-scale turbulence—perhaps Raymond B. Montgomery of the Woods Hole Oceanographic Institution or Hans Panofsky of New York University. Others who could potentially make substantial contributions were Bernhard Haurwitz of MIT, Victor Starr (another Rossby protégé) of the University of Chicago, and Morris Neiburger of UCLA. However, they were already engaged in other research projects. As Rossby ruminated on this list, he feared the models that this group, top-heavy with mathematicians, would choose to pursue. A synoptic meteorologist skilled in *both* descriptive and theoretical approaches had to be added to this mix if the group were to be effective. Indeed, finding people possessing a sense of the atmosphere as well as a mathematical bent soon emerged as a priority in the long-term modeling process. To bring the team together, Rossby recommended that the IAS reach an agreement with a government agency (perhaps the Navy's

Office of Research and Invention) that could supply sufficient funds to allow von Neumann to assemble the proposed group.[38]

Since von Neumann was not concerned with efforts that would aid forecasting in the near term, Rossby also recommended assembling a second group of people from the ranks of the "mathematically, statistically, and synoptically competent and ingenious" in the Weather Bureau, the Air Weather Service, and the Naval Weather Service for the purpose of examining how the ENIAC could be used to compute temporal correlations that would aid day-to-day forecasting.[39] In this way Rossby—theoretician, researcher, and entrepreneur—capitalized on the interest of both Mauchly and von Neumann by proposing projects that would simultaneously tackle meteorological theory and forecasting.

Reichelderfer strongly backed the proposed project even though he was well aware that there was no guarantee of useful results. While preferring that the Weather Bureau should be the governmental organization to take leadership, he was realistic enough to acknowledge that financial constraints might require interdepartmental cooperation. However, he advised Rossby that he would be putting together a program plan with the still-enthusiastic Wexler in the near future, and hoped that Rossby would continue to provide advice.[40]

Rossby saw the Princeton project as a way to advance theoretical meteorology and very much wanted it to move forward. Within a week of his letter to Reichelderfer, Rossby negotiated a tentative contract proposal and funding arrangements with the ORI's staff meteorologist, Lieutenant Commander Daniel F. Rex, and then provided von Neumann with a draft proposal.[41] The proposed research objective was to examine ideas concerning the general circulation of the atmosphere so as to determine its steady-state characteristics and subsequent response to externally applied influences. If sufficient support were forthcoming, the project might even be able to "throw light on the nature of climatic fluctuations." Leaving nothing to chance, Rossby continued with a complete budget description that included numbers and types of people, salaries, travel, and overhead expenses. Noting a lack of suitable candidates for the project, he recommended that Wexler manage the project. In addition to those he had named in his earlier letter to Reichelderfer, he added the name of his protégé Paul Queney, director of the Institut du Globe at the University of Algiers. The proposed starting date for the project was 1 July 1946.[42]

On 8 May 1946, IAS Director Frank Aydelotte signed out a contract proposal to the ORI. However, the research objective had now become "[an] investigation of the theory of dynamic meteorology in order to make it accessible to . . . computing." Specifically, the project proposed to investigate the mechanism and flow pattern of the general atmospheric circulation, the necessity of considering stratospheric as well as tropospheric contributions, the stability of the polar front and other fronts in general, the mechanism and flow pattern of major cyclones including their formation, progress, and stability, and the release mechanisms of local instabilities. However, it did not stop with atmospheric theory or forecasting. The proposal went on to claim that with this computing project, they would take the "first steps towards influencing the weather by rational, human intervention . . . since the effects of any hypothetical intervention will have become calculable."[43] This theme was further emphasized in the Justification Memorandum, which stated that the research program would contribute to achieving the goals of rapidly predicting both short- and long-range weather conditions as well as *controlling* the weather.[44]

In the meantime, with the new machine not yet built (its anticipated completion was 2–3 years away), the project would need to focus on meteorological *theory*. Indeed, meteorological theory could not yet pose problems solvable by this new computational approach. Without this new computer—which could perform calculations 1,000–100,000 times faster than had been previously possible—there had been no motivation to address the relevant theoretical considerations.

The proposal indicated that if enough meteorological personnel could be assembled by fall 1946, these individuals would spend 6 months completing a preliminary analysis of the basic dynamical meteorological problems (those listed above). Following peer review, project members then would determine the most promising research directions. This part of the project would extend until late 1947, at which point the computations could be worked out in parallel with computer development throughout 1948. By 1949, both the machine and the required theoretical work would be complete, and model testing and subsequent modifications would be underway.

Desired personnel for this project would be "first-class younger meteorologists" led by Wexler. Younger meteorologists were required because in the United States they were the ones who had the mathematics and physics background necessary to advance the work. Although older theoretical meteorologists also had these attributes, they were already committed to other academic pursuits. As a complement to this group of young meteorologists, the project would assemble a "prominent group of advisors and consultants" including (besides Rossby) meteorologists and oceanographers such as Harald Sverdrup and Jacob Bjerknes; physicists with radiation and molecular physics expertise such as Edward Teller and Subrahmanyan Chandrasekhar; and an aerodynamicist like Theodore von Kármán. Other experts from a variety of technical fields would round out the number to a total of eight to

ten individuals. With the funding to cover the personnel costs for the meteorological group, the total proposal came to $61,000 per annum.[45] Following
negotiations carried out by Frank Aydelotte, John von Neumann, and Daniel
Rex, the Justification Memorandum was signed on 6 June 1946.

From the very beginning, project supporters had different goals for their
creation. Reichelderfer—who had sparked the project—was interested in
forecasting. Military meteorologists, even those theoretically trained, also
had practical forecasting goals. Rossby and the academics were firmly in the
theoretical camp, although the more intuitive and pragmatic Rossby was
not against applied research. Zworykin and von Neumann wanted a meteorological theory amenable to an attack by computer, an advance that would
ultimately allow for weather control. And the funding source, now called the
Office of Naval Research (ONR), seemed content to support basic research
with the hope of future practical results.

Calls for Manpower and Advice

All major research projects, boiled down to their essence, contain three necessary ingredients: funding, equipment, and manpower. The IAS Meteorology
Project was no exception. The funding was assured by the Navy; von
Neumann's Computer Project team was designing and building the equipment. Manpower availability, however, was in doubt. Meteorologists then (as
now) tended to come in two varieties: theoretical and applied. Theoreticians
had mathematics and physics backgrounds, tended to think in equations,
and viewed the atmosphere as something "out there," not something that
affected daily life. Applied scientists might not be mathematics- and physics-
savvy in a technical way, but they had a sense of the *physical factors* that
influenced the weather. This project needed people who could handle both
theory and applications.

Attracting meteorologists to Princeton was difficult. The more experienced academic meteorologists, i.e., those whose professional careers started
before the war, were primarily theorists. Happily settled on their campuses,
they were not only committed to other projects, they were also extremely
skeptical that the entire computer scheme would work.[46] They were also
under pressure to remain in their current positions due to an overall shortage of theoretical meteorologists, which, combined with the large influx of
recently discharged veterans hitting campuses on the GI Bill, exacerbated
the difficulty of putting faculty members in the classroom.[47] Others were
concerned about the length of time it would take to develop von Neumann's
new computer.[48]

Rossby applied his considerable charm to persuade his handpicked candidates—and sometimes their bosses—that the Meteorology Project was the perfect place for them to use their many talents.[49] Von Neumann obtained commitments from Paul Queney (University of Algiers), Albert Cahn (University of Chicago), and Chaim Pekeris (Columbia University). Wexler would lead the project.[50] The team combination was telling: everyone involved was part of a younger, more mathematically grounded subset of the meteorological community; they were all theorists; and they all had ties to Rossby. Although members of this group had numerous advantages—openness to new ideas being one—their lack of a physical sense of the atmosphere would later prove to be a handicap. With their theoretical bent, it was clear that this project was only going to pursue theoretical development, not the applications desired by the military and Weather Bureau meteorologists.

While others sought people, Reichelderfer sought advice. Sending letters to a number of prominent meteorologists and oceanographers, he described the Meteorology Project as one that would "advance our science materially." The responses were almost universally skeptical of the possible success of the undertaking, based on a realistic assessment of the extant state of meteorological theory in spring 1946. Respondents wondered how the computer could positively influence meteorological questions when there was little understanding of the principles underlying atmospheric behavior.[51] Meteorologists had scant knowledge of the governing equations of the atmosphere. What about the "little known terms"? Should those not be determined first?[52] What were the roles of friction and heat sources and sinks? Would they not need to be determined before developing the equations? Perhaps the computer could play a limited role in solving some dynamical meteorology problems, but nothing more. There would be no solution to the forecasting problem in the near future. And weather control? The experts Reichelderfer consulted all agreed: *that* was an absolute pipe dream.[53]

At least one person had a positive opinion on the proposed computing project: Caltech meteorologist Robert D. Elliott. Elliott had spent some time considering numerical methods while reworking Richardson's World War I-era attempt to forecast by numerical means. At first glance, he thought Zworykin's ideas about using an electronic computer to forecast the weather were "overly optimistic." Upon further reflection, he thought that perhaps these were possibilities overlooked by forecasters desperate to get forecasts out. The increased data density available since the war would make a direct attack on weather prediction feasible.[54]

Ideas were flowing in, and at least a few meteorologists were agreeing to join the project. Von Neumann invited a number of meteorologists to a conference to discuss the project in late August 1946. Wexler, in charge of the agenda, set aside the first two days to discuss scientific questions and the last day to deal with organizational arrangements.[55]

Von Neumann opened the August conference with a discussion of the electronic computer's capabilities. He was followed by Rex, who discussed Navy interests (since it was providing the funding). Rossby then led the discussion of meteorological research. He was followed by meteorologists discussing research topics from their own specialties: Bernhard Haurwitz on dynamics, Hurd Willett on synoptics, Jerome Namias on long-range weather forecasting, and Albert Cahn on the Richardson-Elliott approach to numerical forecasting.

As the conference progressed, topics were prioritized, and assignments made to those who would head up the respective efforts. The first of these were "type problems," i.e., problems not necessarily directly applicable to meteorology, but whose prototype characteristics could be used to test existing or planned numerical techniques and computers. For instance, Pekeris suggested that they address stability questions associated with turbulence— generally a major aspect in thunderstorm-sized systems, but present within low-pressure systems that may be large enough to affect several states at once. One of these was the Heisenberg-Lin equation of stability: a one-variable, linear, total differential equation proper-value problem that could be surveyed for various combinations of parameter values. Non-linear extensions could be added once progress had been made on the easier linear version. The meeting participants decided that Pekeris would start with the Heisenberg-Lin equation when he joined the project in early November.

Meteorological stability problems—such as those involving cooler air overrunning warmer air—were similar to the Heisenberg-Lin problem. These were linearized stability questions superimposed on typical meteorological flow conditions (unlike Heisenberg-Lin, which is superimposed on the Poiseuille flows, i.e., the laminar [uninterrupted] flow of a fluid through a cylindrical tube, including rivers of air). Depending on the circumstances, these problems could be either simpler or more difficult than Heisenberg-Lin. Haurwitz and Panofsky, having ideas for a possible solution, drew this assignment.

The conferees discussed general and specific circulation topics throughout the meeting. They decided to put their efforts toward determining the significance of the stationary and zonally symmetric atmospheric circulation, i.e., which circumstances caused blocking situations that lead to large north-south excursions of air (meridional flow), and which lead to air flowing generally parallel to lines of latitude (zonal flow). Panofsky temporarily received this assignment. Hurricane theory fell to Haurwitz, who had already given it thought and possessed adequate empirical material to make an initial attack.

Meteorologist-turned-mathematician Hunt proposed the analysis of basic meteorological parameters—for instance, velocity and pressure distributions in a large (continental-sized) volume of air. He believed this analysis would help to work out some problems of turbulent flow represented by complex mathematics. All the attendees realized the importance of these very difficult subjects that would require an extensive amount of data and probably use the entire capacity of the planned computer. They agreed that Hunt should work on turbulence.

The meteorologists also discussed the continuation of Elliott's efforts to renew Richardson's attempts to directly integrate the equations of motion. Since the computing machines available to the group were considerably more advanced than those available to either Richardson or Elliott, they agreed that the direct numerical attack should be repeated immediately. They were not sure whether continued efforts should be made to eliminate the flow velocities from the equations, since that approach seemed to lead to analytical difficulties and questionable approximations.

At a final evening meeting of the working group, members discussed assignments and the role that each would play. Montgomery would serve as a part-time consultant, and give his attention to numerical forecasting by direct integration. Walter Elsasser, fully occupied at the RCA laboratories, would consult when available. Cahn's task would be to undertake the Richardson calculations by formulating the dynamic equations and setting them up for a numerical approximation method. He would become familiar with the ENIAC, and possibly with Harvard's Mark I computer, and then supervise the actual integration. Panofsky and Haurwitz would assist with the equations, von Neumann would address the numerical approximation, and von Neumann (together with mathematician Herman H. Goldstine, the assistant director of the Computer Project at the IAS), would deal with the computing machines themselves. Hunt, who had resigned his Army commission, was scheduled to remain in Princeton until his discharge on 1 November, at which time he would decide whether or not to remain with the project. In the meantime, he would assist Cahn and also work on general atmospheric circulation. Panofsky wanted to remain involved with both NYU and the Princeton group. Haurwitz, at short-handed MIT, was unable to make a full-time commitment until February 1947. He intended to concentrate on hurricanes and instability. Wexler would remain at Weather Bureau headquarters until late 1946 and then move to Princeton once housing became available. He planned to split his time between Washington and Princeton, where he would supervise the working group. In the meantime, he would make frequent visits to Princeton. Von Neumann would spend the last half of September at Los Alamos, but would remain in Princeton after 1 October. Those associated with the project decided to hold periodic meetings to discuss problems and assess their progress.[56]

Despite all the time and effort that went into this meeting, very little would come of it. The members were physically separated and occupied with other projects. What could have been a jump-start for the Meteorology Project turned out, as described below, to be a false start. This was hardly apparent at the time. Reporting to bureau chief Reichelderfer, Wexler shared his conviction that the "abrupt discontinuity in speed" represented by the new computing machine would make a substantial difference in the discoveries of theoretical meteorology by reducing calculation times. Despite its theoretical bent, Reichelderfer remained steadfast in his conviction that the project was of the greatest importance and must be kept moving forward.[57]

By fall 1946, in any event, the Meteorology Project had established its priorities and arranged for team members. Reichelderfer and Wexler, having thrown their complete support behind it, were clearly enthusiastic. But the project's theoretical ambitions were worrisome for the weather services that had little use for meteorological theory. What they needed was a way to get better forecasts out faster, using less subjective techniques.

The Meteorology Project Takes Form

The original proposal had called for a fairly large group of collaborators to work on the Meteorology Project. A drastic shortage of housing in the Princeton area quickly derailed this plan, preventing several investigators from joining the group and subsequently becoming unavailable or less available than they had been. Indeed, the combination of scarce housing and a lack of investigators to supervise soon convinced Harry Wexler to remain in Washington and periodically commute to Princeton.

Visiting in mid October, Wexler found Paul Queney, Gilbert Hunt, and Albert Cahn on site. Von Neumann was very pleased with the progress being made on the general circulation model and the setting up of the Richardson equations. He thought—rather over-optimistically, as it turned out—they would be able to put the equations on the underutilized ENIAC in the near future. However, the group was being hampered by a lack of office and living space, and consequently, of personnel. Wexler thought the group needed coddling to make sure it got off on the right foot. He was providing both advice and meteorological information to these theoreticians so that they would have evidence of actual atmospheric behavior. Included in these data were wartime-prepared historical upper-air reports, a requirement for atmospheric modeling.[58] This happy situation was extremely short-lived. Returning to Princeton a mere two weeks later, Wexler was shaken to discover that von Neumann was ready to abandon the project. Living and working conditions had created an unstable personnel situation.[59] The computer work had started in the IAS's Fuld Hall boiler room in June 1946, but other Institute members, finding little merit in the venture, wanted it even more "out of sight and out of mind."[60] The temporary building being moved in to house the Computer and Meteorology Projects was not yet ready, so office space was simply unavailable. Likewise, Depression-era Works Progress Administration housing being hauled to Princeton to provide living quarters had yet to be installed.[61] "We must not allow this important project to lapse," a troubled Reichelderfer jotted in the margin of Wexler's report.[62] Indeed, the project's progress report for the period ending mid November commented that the larger group originally anticipated to compose the project could not be assembled due to the housing shortage. They would form a smaller group instead and attempt to create a more cohesive unit.[63]

In mid November a group of prominent meteorologists and oceanographers visited von Neumann in Princeton to share their ideas about the use of the computer and approaches to solving the numerical forecasting problem. Computer forecasts that produced abnormally large changes in pressure tendencies, already seen by Lewis Fry Richardson, Robert Elliott, and an AAF officer at UCLA (Lieutenant Philip D. Thompson) in their work, indicated that both a new mathematical approach and changes in the observational network were needed. Von Neumann thought that they would need to make trial runs on the ENIAC to ensure they were ready when his new computer came on line. The personnel onboard had been significantly reduced: Hunt was temporarily out of town, and the frequently absent Cahn had been fired. However, the visiting scientists had lifted von Neumann's spirits: with Chaim Pekeris and Queney settling in, he was no longer talking about leaving the project.[64]

Wexler's next report (as ever, neatly typed with single spacing and tight margins) noted that von Neumann was pushing for an objective method of determining the field of divergence, i.e., any area where air molecules are moving apart from each other, in the atmosphere. MIT and NYU, he complained, continued to use unacceptable subjective, non-mathematical methods. In order to use a more objective method, the project would need very good upper-air data—temperature, pressure, and wind velocities aloft. NYU possessed such data sets, and would pick one as a case study. Since the required calculations were extensive, they decided to ask the Bureau of Standards for assistance. Fortunately it was happy to help.[65]

Maintaining the momentum building in postwar research, Carl-Gustav Rossby organized a conference at the University of Chicago to discuss the most important topics in meteorological research. The 18 conferees at the December 1946 meeting were primarily theoretical meteorologists from the University of Chicago, but meteorologists from NYU, MIT, the Meteorology Project, and the Weather Bureau attended too. Looking back at the list, two striking anomalies emerge. There were no representatives from the West Coast universities—UCLA and Caltech. And a final attendee was from the Soviet Union: Commander Ryshkov of the Hydrometeorological Service. Caltech's absence is not a surprise. Robert Millikan's meteorology program, focused on creating entrepreneurs in operational meteorology who could provide contract services to industry, was not the least bit theoretical. Soviet Commander Ryshkov's participation in late 1946 demonstrates the continued free flow of scientific ideas in the immediate postwar period.

The assembled meteorologists considered five major topics: the relationship of the wind and pressure fields, the scales and types of atmospheric perturbations, stability criteria, general circulation, and surface waves of finite amplitude. They held detailed discussions on the current and developing meteorological theory in each of these areas.[66] On the important topic of model development, Rossby argued that model conception needed a requisite physical nature in to be useful. The most fruitful path would be to create simple dynamical models that characterized the atmosphere.[67]

::

Early winter had also heralded the arrival in Princeton of Army Air Force Lieutenant Philip D. Thompson. A man of extraordinary intelligence and unbridled ambition, Thompson had received wartime meteorology training at the University of Chicago, but was subsequently assigned to air traffic controller training. Thompson was not happy to be stuck in air traffic controller school after finishing an intensive course in meteorology. He was subsequently relieved of those duties because of a temperament "not suited to the high nervous tension . . . developed during this duty."[68] Thus, when Thompson requested reassignment to duty as a weather reconnaissance observer, he was almost denied that position for temperament too.[69] Ultimately, he was returned to the Air Weather Service and assigned to the Army Weather Station, Long Beach Army Air Field, California. While there, he learned that the Army Research Weather Station at UCLA needed a few more officers on its staff. Thompson applied, noting that his "greatest interest, and, in consequence, usefulness lies in meteorological research, rather than in operational forecasting."[70] The request approved, Thompson made his move into meteorological research.

Working on objective forecasting techniques, Thompson's interest was piqued when he heard about the work of the Meteorology Project from the UCLA-based Holmboe. Although Thompson later claimed Holmboe had shared a *New York Times Sunday Magazine* interview of von Neumann and Zworykin with him in the fall 1946, no such article exists.[71] What is clear is that Thompson placed a call directly to his commander, General Ben Holzman, and convinced him to authorize a trip to Princeton to visit with von Neumann. The latter was sufficiently impressed to invite Thompson on board and the Air Weather Service assigned him to serve a tour of duty with the Meteorology Project.[72] This was an extraordinary event: at the time Thompson was only a first lieutenant (a very junior officer) with a bachelor's degree from the University of Chicago. First lieutenants do not have direct pipelines to generals. That Thompson had the chutzpah to call Holzman directly, much less request an audience with von Neumann, is only a small hint as to the measures he would take to get what he wanted, when he wanted it. While this trait generally worked to advance his career, it would not always prove endearing to his colleagues.

With the arrival of Thompson at the Meteorology Project, the Air Weather Service made its first contact with numerical weather prediction. Equally significant, the military weather services had a member on site. For the first time all three constituencies—academic meteorologists, Weather Bureau civilian meteorologists, and military meteorologists—were represented.[73] By accepting a military member, the project—perhaps inadvertently—opened its work up to military scrutiny in an unintended way. Although von Neumann's ONR contract directed him to submit periodic formal reports, he enjoyed complete control over their content. Von Neumann could not, however, control the content of Thompson's reports. Thompson would be able to report directly to his military superiors without being censored by von Neumann. Indeed, the nature of what kinds of statements were released about the project were about to become a point of contention.

The Air Force's seeming inability to allow its personnel to fill positions without having an organizational designation required Thompson to be assigned as the Officer in Charge (and, literally, sole member) of the Seventy-Second AAFBU Detachment (Special Projects Unit). As such, he needed to create a mission statement for himself and regularly report on his activities. Despite being on the job for only a few weeks, Thompson filed his first report (classified "Restricted") at the end of December 1946. His detachment's mission: to restate meteorological problems as hydrodynamical problems, to formulate them as mathematical problems, and to solve meteorological problems capable of physical analysis. To do this, he would need to coordinate information from meteorology, fluid mechanics, mathematics, and electrical engineering. Therefore, Thompson had reviewed the work of Richardson and Elliott, who had used finite differences to examine the underlying mathematical structure of the graphical methods promoted by Vilhelm Bjerknes. However, finite differencing was not a viable approach because available data were not sufficiently representative. Therefore, it would be necessary to "examine systems which have simple analytical form, but which may be identified with the real atmosphere."[74] Since the Harvard Computation Laboratory and the Naval Ordnance Laboratory were also working on a numerical solution to the hydrodynamical problem, even if in a more general sense that the Princeton group, Thompson noted that it would be important to stay in touch with them.[75] Because of the heavy military presence and mission at these labs, Thompson was probably more comfortable with this arrangement than were Pekeris or Queney.

Thompson had only been in Princeton a short time when he heard from a former UCLA acquaintance and wartime-trained meteorologist much interested in the Meteorology Project: Jule Gregory Charney. Charney had completed his undergraduate work in mathematics and physics (Phi Beta Kappa, 1938) at UCLA, and was a mathematics doctoral student there when he first heard Holmboe lecture on fluid turbulence. Attracted by the subject matter, he accepted an offer to become Holmboe's assistant and simultaneously participate in the meteorology program being established in support of the war effort. And so Charney made the switch to meteorology. More comfortable with mathematical explanations than the more descriptive techniques then the mainstay of meteorological studies, Charney went his own way, producing a masterful PhD dissertation (later to be published in the *Journal of Meteorology*) titled The Dynamics of Long Waves in a Baroclinic Westerly Current.[76] Having completed his degree in 1946, he was awarded a National Research Council fellowship and set his sights on the University of Oslo. There he would study with the leading mathematician of the Bergen School, Halvor Solberg. However, en route he stopped off at Chicago and called on Rossby, who encouraged him to stay for awhile. Charney remained at Chicago for almost a year, taking part in free-wheeling discussions of meteorological theory with the Chicago staff and the many foreign visitors who passed through. Given Charney's mathematical approach to meteorology, Rossby had taken him along to Princeton for the August 1946 meeting that set up the Meteorology Project. While there, meeting with von Neumann, Charney was drawn to the ideas surrounding numerical weather prediction. Now preparing, at long last, to leave Chicago for Oslo in early 1947, he wanted solid first-hand information from Thompson on what was happening at the project.

Charney had been mulling the problem of numerical weather prediction since the previous summer. He wanted to share his ideas with Thompson about how to consider wave speed and other motion ideas, with an eye to setting up a system of solvable equations for atmospheric circulation. Since it would be hard to share these ideas by mail, Charney suggested that Thompson invite him to Princeton for a visit since they had "all that Navy money lying around."[77] This gambit worked: Charney was able to make the trip in March just before leaving the United States, and he dropped his detailed findings in the mail to an eagerly waiting Rossby.

The news Charney had to share was not promising. In his opinion, the Meteorology Project was the "ugly duckling" of meteorology, but had the potential to become a "swan." Charney had several concerns about the Princeton project, not the least of which was that meteorology was the "weak sister." With no cooperating meteorology department, the meteorologists were largely isolated, and worse, seemed to have no coordinated approach. Queney—with limited facility in English and little rapport with von Neumann—was working on a variety of wave motions, none of which seemed to have any relation to the real atmosphere. Pekeris and Panofsky (working at NYU) had almost nothing to do with the project. Thompson was very capable, Charney believed, but had little knowledge of other work. Thompson was, however, the only person on the project who realized the significance of rapid readjustment processes for large-scale motion in the atmosphere. Von Neumann regarded scale questions as being of secondary importance and attributed instability in the calculations to the computation processes themselves and not to the physics of motion. This concerned Charney. He thought project members needed to consider the possibility of dynamic instabilities, which could lead to computational errors, being inherent in the system. In his opinion, the instability to which von Neumann alluded was the same phenomenon already discovered by Victor Starr when he was "playing around" with difference equations at the University of Chicago. Charney thought they would all be better off if the Meteorology Project were co-located with Rossby in Chicago.[78] However beneficial that might have been, this did not happen. Indeed Paul Queney eventually bolted, leaving Thompson on his own for almost a year.

In March 1947, Philip Thompson wrote a survey of the IAS project, primarily for Air Weather Service consumption, but also with the idea that the project needed to get out some accurate information about its work to counteract sensational publicity appearing in the popular press. (Sample headline: "Scientists Get Ready to Do Something About the Weather; World-Wide Observation Planned; Force to 'Counter-Attack' Storms Considered," from the 20 January 1947 *Chicago Sun*.) He worried about overselling the project and expressed the hope that readers of his survey would do so without "undue optimism, though certainly not with preconceived pessimism." He was aware that many forecasters were very skeptical about numerical techniques. Results produced in haste could ultimately lead to dwindling support for this new approach to weather prediction. Thompson went on to explain the motivation behind numerical weather prediction. The hydrodynamical equations that describe the physics of the atmosphere could not be solved analytically. That being the case, a new line of attack had to be undertaken. With the introduction of the new high-speed electronic computers and increasing data availability, numerical analysis techniques could be applied to the problem. Thompson explained these methods in great detail and sent off copies of his report to the Air Weather Service for internal distribution only.[79]

More than anything else, Thompson wanted to get out some corrective publicity before the meteorological community looked upon the Meteorology Project as an unprofessional excursion into scientific hype. He sent a copy of the report to fellow AAF officer Robert Bundgaard with a note saying the survey was deliberately "conservative and vague" because von Neumann wished to publish his own paper on computational stability.[80] Wexler, having received a copy also, agreed with Thompson that the report should be published just so they could remove a "good deal of the mystery" surrounding the project.[81] However, having been scorched by inaccurate press reports, von Neumann did not want Thompson's survey published in a peer-reviewed (or any other) journal, so ultimately it was published only in restricted Air Force technical reports.[82] Consequently, there was little official information about the project reaching the scientific community or even the Weather Bureau staff. Indeed, after receiving another trip report from Wexler about a recent visit to Princeton, Reichelderfer wrote at the bottom of the memorandum: "This project is still in the 'prospecting stage' but it represents a possibility which has general interest and perhaps our field service should be informed. What do you think of a brief, factual (not visionary or over-optimistic) item [for the Weather Bureau newsletter]?"[83]

Thompson was not the only participant to pen a report on numerical weather prediction. Albert Cahn, fired from the Meteorology Project, had subsequently joined the National Bureau of Standards. In June 1947, he sent his report on numerical weather prediction to Wexler. Cahn noted that insufficient data density and poor data accuracy presented substantial obstacles. Indeed, in 1922 Richardson had named these same data deficiencies as the ultimate source of the errors in his prediction. This had led to what Cahn labeled "a symbiotic inertia of form: there is no use developing methods to get extensive, accurate observations since there is no use for them; on the other hand, there is no point in funding computing techniques to do numerical forecasting since the required observations are not available." However, with new computers being developed, the balance had changed. Now they needed to determine which observations were truly prerequisite to good forecast output from models. After considering a number of questions that would guide the answer on observations, Cahn asked, "Do you think the problem of predicting the weather is worth all the effort we seem to be spending?"[84] Cahn was a theoretician, not an operational meteorologist providing daily forecasts, and his doubts deeply troubled Weather Bureau leaders. How many other theoreticians felt as Cahn? How many forecasters, skeptical that numerical methods would ever work, would agree with him?

After Paul Queney's September 1947 departure, Thompson was the only person remaining in Princeton on the Meteorology Project "team." In his project report, Thompson noted that he was continuing his mathematical-physical research, while the NYU subcontract group under Hans Panofsky was doing synoptic and empirical work. Cooperating government agencies were doing the extensive numerical computations.[85] Reporting to Francis Reichelderfer, Wexler was impressed that Thompson had managed to produce significant results on his own despite chronic staffing shortages. The Weather Bureau was contributing by giving advice and suggesting problems to be solved. Wexler was addressing a variety of theoretical topics, and the bureau staff was providing hand-drawn analyses in support of the numerical weather forecasting project. Reichelderfer was very supportive and very much desirous of keeping the Meteorology Project on track.[86] However, the bureau had a shortage of analysts and could offer only limited help. The Meteorology Project had no analysts at all.[87] This would prove to be an ongoing difficulty. The only way to secure data to use for the models' initial conditions was from an analyzed chart. Why? Because when raw data came in, they were plotted on a chart. The chart was then analyzed with the familiar lines of equal temperature, pressure, wind speed, etc. Analysts then placed a grid over the analyzed chart and extrapolated data values to the grid points. These extrapolated values became the initial values (or initial conditions) for the calculations. No analysts—no initial conditions. Therefore, the project had to have access to enough qualified analysts to provide starting point data as well as analyzed charts to verify computer predictions.

Sharing Reichelderfer's desire to keep the project moving ahead was Jule Charney. Since spring 1947 Charney had been at the Oslo Institute of

Meteorology, working to develop a solvable set of equations out of the basic hydrodynamic equations. For him, weather forecasting was primarily a computing problem that required "one intelligent machine and a few mathematico-meteorological oilers." Charney thought the Meteorology Project soon would have von Neumann's computer, but it was lacking in "oilers." For that reason—and because he was convinced that he had found a solution to the forecasting equations by applying filters to get rid of unnecessary "noise"— Charney wanted very much to join the Princeton team instead of accepting offers from the University of Chicago and UCLA.[88] His eagerness to try out his new filtering method on the computer was not his only reason for wanting to join the Meteorology Project. Charney's misgivings about the project's direction spurred his desire to go to Princeton. "Unless some physical ideas are brought to bear," Charney confided to a close colleague, "the project will die out through mathematical sterility. I have no delusions of grandeur about my own possible contributions, but at least I may help to give it the right slant."[89]

Von Neumann was delighted to hear from Thompson that Jule Charney wanted to join the team. He invited Charney to become a project member at the conclusion of his fellowship year in Norway and inquired of his requirements. He also wanted to know if Norwegian meteorologist Arnt Eliassen, working with Charney in Oslo, would be willing to come too.[90]

Accepting von Neumann's invitation in early 1948, Charney used the occasion to offer his opinion on a suitable approach to numerical weather prediction and advise von Neumann of his progress. Charney had determined that the dynamical equations of the atmosphere could be reduced to a single linear partial differential equation in the pressure and of the first order in the time, by assuming that large-scale atmospheric motion is governed by the conservation laws of entropy and potential vorticity, and by conditions of quasi-hydrostatic and quasi-geostrophic equilibrium. (Assuming the system is quasi-hydrostatic means that air parcels experience only extremely small accelerations in the vertical. Assuming it is quasi-geostrophic means that the actual horizontal wind can be replaced by the geostrophic wind—the wind resulting when the Coriolis force due to earth's rotation balances the horizontal pressure force. It also allows for the neglect of the vertical advection of momentum. This assumption works for large-scale systems outside of the tropics and away from frontal systems.) If the assumptions held, they would just need the easily obtainable initial pressure distribution to integrate the equation. If the short wave motion could not be eliminated from the dynamical equations, they would need to start with the initial vertical velocity and horizontal divergence, or the initial pressure tendency—quantities that were not determinable with sufficient accuracy for numerical techniques.

Charney also noted that there was no rule to distinguish large- and small-scale motions, but that his filtering scheme removed all wave motion smaller than those small-scale wave cyclones (extending several hundred miles) that appeared on weather maps. Furthermore, such a separation was not strictly mathematically justifiable, since the equations were not linear. However, it turned out the small-scale perturbations could be safely ignored in the first approximation, thus making the equations easier to solve. Therefore, although the transition motions might indeed be meteorologically significant, there were not sufficiently accurate data to handle them. Charney felt they could hope to "forecast the principle [*sic*] large-scale current systems" and then regard these as steering currents for the smaller scale motions.

Having dispatched the troublesome small-scale motions, Charney set his sights on long waves. Pointing out that no one understood their controlling mechanism, he thought they should forecast large-scale perturbation movements for one to three days and see what happened. If they could accomplish that, then the money and effort would be worthwhile.

Turning to more mathematical considerations, Charney referred back to Richardson's efforts. He declared the importance of looking at computational stability and how it corresponded to grid spacing and temporal scales. Because the atmosphere is a dispersive medium, they had to account for the influence of energy propagation from outside the area of interest. This outside energy does not propagate at the same speed as the disturbances themselves and as far as Charney knew, "the question of energy propagation in finite amplitude systems is unsolved." With meteorologists in Chicago, Oslo, and Stockholm looking at these perturbations, they had found "many examples" where the "influences of neighboring atmospheric perturbation on one another" were propagated faster than the disturbances themselves and even faster than the wind velocity. Models could help in the investigation, but selecting those models was a physical problem and therefore numerical forecasting would require a combined effort of mathematicians and physicists. But perhaps most importantly, Charney realized that not even a mathematical and physical approach would be sufficient. The meteorology group needed people who knew enough about meteorology to know "when and how to make the approximations." This was an extremely important insight. The equations defining atmospheric motion were never going to be solved if all of the terms were left in. Therefore, the terms least likely to affect the solution would need to be removed or a value substituted in for them. Practicing meteorologists already made those approximations in their heads during the course of the forecasting day. Someone who had been forced to make tough decisions about what to keep and what to throw out would be crucial to the success of this project. To this end, Charney recommended bringing over Arnt Eliassen from the Norwegian Meteorological Institute. Eliassen possessed experience in both *synoptic* and *theoretical* meteorology, and was also interested in numerical solutions. Sverre Petterssen—head of the Norwegian Forecasting Service, who had spent the war years advising the Meteorological Office of the British Air Ministry (and had participated in the D-Day invasion forecast)—had approved a year's leave of absence for Eliassen, who was quite willing to come. Charney thought it was important to combine theoretical work with empirical data, and therefore that it was important that some members of the group have "intimate experience with actual weather processes." For Charney, Eliassen was that man. To make sure Eliassen was "up to speed" on the latest synoptic work, Charney had already arranged for him to visit with Rossby at Chicago and go to Weather Bureau headquarters.[91] In retrospect this was a particularly significant development, marking the emergence of the international Scandinavian-influenced "Tag Team" approach to developing numerical weather prediction.

Von Neumann, responding to Charney in early February, said that he was "very anxious" to have him and Eliassen with the project "next [academic] year." In response to Charney's inquiry regarding "professional and sub-professional help," he reiterated who would be locally in place: Philip Thompson and Albert Hunt. Hans Panofsky and Bernard Haurwitz were collaborating from NYU, and Harry Wexler would continue to provide assistance from the Weather Bureau.[92] This meant the "group" would consist of Charney, Eliassen, Thompson, and Hunt—a significantly smaller group than had been foreseen in the initial plan set forth by von Neumann, and even small by his revised plan.

Nonetheless, the Meteorology Project was reaching a minimum staffing level. Charney was very pleased to be joining the Princeton team and bringing Eliassen with him. Charney's only concern was the salary, which was less than he had proposed. One reason for the concern: he viewed the Princeton job as temporary and anticipated that he would be moving his family from Los Angeles (where they had left their household goods while in Norway) to Princeton and back to Los Angeles within a year. Apparently Charney thought that his contribution would be over fairly quickly so he could rejoin UCLA's Meteorology Department.[93] Indeed, Jacob Bjerknes of UCLA had asked Charney to let him know when he would be available after his "well accomplished job in Princeton."[94] Charney, as it turned out, was to be in Princeton much longer than anyone anticipated.

The End of the Slow Start

Despite the discussions of numerical forecasting, the project was still primarily concerned with atmospheric theory in spring 1948. Philip Thompson, worried about his vulnerability to transfer because the end of his assignment was approaching, made clear to his Air Force contacts that he wanted to remain with the project until he was sure the fundamental problems related to developing equations that described only essential atmospheric phenomena were in place. Once that was done, numerical prediction would be within the group's reach. In his opinion, the Meteorology Project was the "first, and at present only, direct and potentially successful approach to fundamental problems." He was completely convinced that the project could succeed. Furthermore, Thompson anticipated that funding from Navy contracts would continue indefinitely, although it appeared that the scope would be widened to include a broader range of geophysics topics.[95] This last comment probably referred to the request made by the ONR's Commander Roger Revelle to expand the project's geophysics base.[96]

Little of substance had been accomplished during the first 2 years of the project, but with the personnel situation about to improve, it was poised to take advantage of a new talent mix. Hunt, having recently completed his doctoral program, rejoined the project. That meant Thompson and Hunt were the only team members in Princeton. Weather Bureau headquarters personnel handled plotting, analysis, and data preparation as time permitted. Since there were no funds available to do this work, bureau employees could only work on Meteorology Project tasks once the day's routine work had been completed. With von Neumann's computer still under construction, the meteorologists anticipated using either the Bureau of Standards' computer or the ENIAC to perform the initial calculations.[97] Yet as the progress report for the six-month period ending in mid May of 1948 indicated, there was still no organized and established meteorological theory.[98] Without such a theory, they would be unable to move toward the prediction phase. But help was on the way. By mid-summer Jule Charney and Arnt Eliassen would arrive from Norway, and things would start to look up for the hard-luck Meteorology Project and for the prospects of numerical weather prediction.

5 :: An International Atmosphere: Carl-Gustav Rossby and the Scandinavian Connection (1948–1950)

With the arrival in Princeton in August 1948 of Arnt Eliassen—the first member of the Scandinavian Tag Team—an international atmosphere returned once again to the Meteorology Project. By then, Chaim Pekeris had moved to Israel and Paul Queney had returned to France. Both Pekeris and Queney, however, had been working *in* the United States before being asked to join the project. Eliassen, in contrast, was an *imported* scientist—imported to provide some measure of atmospheric reality to a heavily theoretical project. Indeed, he was imported because Carl-Gustav Rossby, the de facto head of the Meteorology Project, was determined to have a significant influence on its outcome despite being an ocean away in Stockholm.

If this group were focused on the development of meteorological theory, why did it need personnel with *synoptic* experience? Synoptic meteorology relies on data collected worldwide and analyzed locally to make predictions. It was a very subjective endeavor, and theory-based dynamicists considered it to be more art than science. However, Rossby recognized that any theory used as a basis for a computational solution had to include those factors that were either consciously or unconsciously used by forecasters who were using significant skill to turn raw data into a representation of the atmosphere from which they could make a prediction. Any additional variables could be added once the field forecasters' approximations and assumptions had been included.[1] Therein lay a potential pitfall for this still (after 2 years of existence) fledgling group—a point touching on the fundamental reality of modeling. If the team members were looking strictly at elegant numerical solutions to the hydrodynamical equations, then they could develop internally consistent models. Such models could produce forecasts for conditions at multiple atmospheric levels that were correctly correlated with each other, but not necessarily have any relation to atmospheric reality—that is, to nature itself. As Rossby noted, the equations needed to be viewed as tools for studying problems suggested by the atmosphere, not as an end in themselves.[2] Without solid synoptic support, Charney's fear that the group might become mathematically sterile would come to pass. Rossby's mission was to put that fear to rest.

Rossby's Off-Site, On-Scene Research School

Charney respected Rossby tremendously and looked to him for guidance and intellectual stimulation. As Charney put it days before his departure to Norway: "You will see me in Sweden if I have to ski there from Oslo."[3] Thus, once Charney joined the project, Rossby had a free pass to influence it, and he took every advantage of the opportunity. As Rossby was shuttling across the Atlantic while organizing the University of Stockholm's new meteorology department, Charney kept him apprised of the project's progress.[4] In return, Rossby provided a steady stream of ideas, personnel, and encouragement to Charney. Rossby also provided a publication venue: *Tellus,* the new geophysical journal he had founded and was editing.[5]

Tellus was not, Rossby pointed out, "just another journal." As the editor, Rossby expected his new journal to be truly international (despite the Cold War), to mirror what was happening in geophysics, and to serve as a bridge between the various branches of the geophysical sciences. Since its international aspect was essential to spreading new meteorological ideas around the globe, this remained Rossby's greatest concern. Indeed, *Tellus* was specifically created to advertise research already published in other journals including the American Meteorological Society's *Journal of Meteorology*—also founded by Rossby. Whereas the *Journal* had been created to provide a place to publish peer-reviewed, meteorological research venue for readers in the United States, *Tellus* provided a peer-reviewed journal for geophysicists throughout the world and accepted contributions written in English, German, and French.[6]

These developments point to the ways in which Rossby personally influenced the paths that meteorology in particular, and geophysics more generally, would take in the middle years of the twentieth century. In short, they underscore the importance of Rossby's far-flung, loose-knit organization as a research school.[7] An internationally respected meteorological researcher by the late 1930s, Rossby had previously established two meteorology departments in the United States (one at MIT and one at the University of Chicago) and was beginning another in Sweden as the Meteorology Project was gaining momentum. He had also directed research programs for the US Weather Bureau and been an active participant on the US government's Research and Development Board. Notoriously lacking in attention to administrative details, he was happiest setting up programs, arranging funding, and letting others take care of the day-to-day operations.[8] He was not hindered in getting what he wanted no matter where he worked—he had the ear of people in high places and worked those connections.

Rossby's research focus changed with the times; he worked on whatever presented the most scientific potential. From the 1920s through the 1950s, Rossby would move from aviation-related concerns to dynamics, on to numerical weather prediction, and later to tracking radioactive isotopes as a way of determining the general circulation of the atmosphere. In each case, he was at the cutting edge of a new field in the atmospheric sciences.[9] He drew students from all over the world to wherever he was and continually pushed them and his colleagues to publish their work and publish it quickly—in *Tellus* or other appropriate journals.[10] He was responsible for finding the right people for the right job—keeping up with a large number of correspondents (as existing archival evidence demonstrates). Rossby's vast network of contacts and his influence over them would prove very important to the Meteorology Project's success.

Moving to a New Level

The summer of 1948 saw a major infusion of enthusiasm and meteorological insight, culminating when Charney and Eliassen joined the Meteorology Project. Charney came armed with, and ready to try out, the techniques he had developed for filtering out noise, e.g., sound and gravity waves that complicated the physical solution to atmospheric motions, but did not in practice influence the weather. Eliassen, with his well-rounded combination of synoptic, theoretical, and numerical skills, offered the promise of practical atmospheric experience to counterbalance the heavy theoretical emphasis. From this point on, the project maintained much closer ties with Rossby and, not coincidentally, made rapid progress toward a formal theory.

Before Charney's arrival, team members had individually taken on various problems to solve without first determining where they needed to go and how their individual projects might take them there. Once Charney took over, the emphasis of the four-man team shifted to mapping out a path, and then identifying and solving more general groups of problems along it. Charney, Eliassen, Hunt, and new arrival John C. Freeman focused on developing a method to mathematically integrate the meteorological equations so they could be solved by the new IAS computer. To help them reach their ultimate goal, they set up intermediate goals: finding the governing laws of atmospheric motion, finding a way to numerically integrate those laws when written as differential equations, and finding the data requirements needed for a solution. To address these goals, the group proposed to consider a "hierarchy of 'pilot problems,'" each of which would contain more physical, numerical, and observational aspects of the general forecast than the preceding one.

Rossby had already shown that planetary circulations of the atmosphere were more amenable to quantitative techniques than for more localized regions, so the team members decided to start there: more was known about large-scale than about small-scale motions.[11] They could address this large-scale motion by using the hydrodynamical equations for a non-viscous, adiabatic fluid (one that does not exchange heat with its surroundings). As discrepancies appeared between the numerical solution and the atmosphere's observed state, they could modify the equations by adding one parameter at a time to ascertain its effect. In this way, they hoped to avoid the frustrations that Richardson had faced in the late teens and early 1920s as a result of trying to do "too much, too soon."

The hydrodynamical equations were troublesome because they governed atmospheric motions—including sound and gravity—that were of no consequence to the meteorological situation. Therefore, the team had to filter out these smaller-scale motions so as not to obscure the larger-scale motions.[12]

Because failure to get it right would doom the rest of the project's efforts, Charney dealt with noise first. Just a few weeks after his arrival in Princeton, Charney laid out his ideas on noise and other topics in a long, detailed, technical letter to von Neumann, who was spending the summer working at Los Alamos. Charney wrote that the so-called primitive equations (those used by Richardson) were not going to work because there was no method of accurately measuring horizontal acceleration and divergence—both of which were very small differences between very large terms. Therefore, the noise level in smaller-scale motions would mask the larger-scale components. No matter how much observational techniques improved (and they were unlikely to improve *that* much), the difficulties associated with noise would persist. Since the horizontal divergence term appeared in both the continuity and vorticity equations, Charney chose to eliminate it by combining those two equations. The still unobservable horizontal acceleration term would remain, but Charney argued that it could be replaced by the geostrophic approximation (where the Coriolis force is equal and opposite to the pressure gradient force) that would filter out the gravity waves. If the gravity waves were included, large initial data sets would be necessary to prevent an unstable computation. Using the filter would reduce the size of the required initial data sets.[13]

Charney acknowledged that two methods had already been proposed for solving the resulting system of equations: one by Philip Thompson and one by John von Neumann himself. Thompson's required one to guess the value of certain derivatives, creating a situation where the solution would be neither stable nor converge. Von Neumann proposed that the kinematic boundary condition be used to determine the surface pressure change. However, this approach required solving a three-dimensional equation as a two-dimensional one. Charney disputed that approach. Instead, he proposed his own method that would be a direct integration for pressure by replacing the space derivative with finite differences. Assuming that the starting equations were correct, both horizontal and vertical influences would propagate at a finite rate. One would then only need to possess initial data in a finite region around the forecast point of interest.

Studies of pilot models showed that the mechanisms of both horizontal and vertical propagation were very similar. Charney's proposal for an "immediate attack" on numerical forecasting was to describe the initial pressure field in such a way that the average motion was defined as being two-dimensional (even though the atmosphere is a three-dimensional space) and replace the actual atmosphere with a barotropic atmosphere. In a barotropic atmosphere, surfaces of constant density or temperature coincide with surfaces of constant pressure. Thus, for any given pressure level, the analyzed lines of equal temperature and equal height would be superimposed.[14] The continued study in two-dimensions would provide needed practice and experience to prepare for the eventual three-dimensional approach. Since the two-dimensional model would be less difficult, the team would be able to uncover modeling mistakes and data errors more quickly. And no less important on a project this large, reaching an intermediate goal would provide a distinct psychological boost to the team members.[15] It was unrealistic to expect that they could model the atmosphere successfully on the first try. But if the team members could get a simple form of the model to work, they could build on that success.

When von Neumann inquired about Thompson's approach, Charney restated his contention that Thompson's iterative method would *amplify*, not eliminate, the noise. The atmosphere is, after all, three- and not two-dimensional, and must be treated as such. He also explained again that there was no other option than to eliminate the divergence term from the equations because it could not, under any circumstances, be measured.[16] So while von Neumann clearly had the upper hand when it came to computer design and numerical analysis techniques, Charney had the superior knowledge of atmospheric processes.

Charney also had the ear, and the support, of Rossby. Settled in to the project's routine, Charney wrote to his mentor, apprising Rossby of the latest developments and including a copy of his letter to von Neumann. Charney was discouraged by their slow progress: von Neumann was out of town, and Eliassen was distracted by his search for housing. However, Charney was happy to report that the objective analysis part of the project, which had been underway at NYU, was being dropped, along with its requirements for the bulk of the funding. But what Charney really wanted were Rossby's comments on the ideas he had presented to von Neumann.[17]

With his acolyte on board in Princeton, Rossby had in Charney a built-in conduit to influence the project's direction. His correspondence with Charney vividly illuminated Rossby's style of mentoring. He immediately started filling the pipeline with technical and professional advice. But instead of providing the feedback Charney had requested, Rossby launched into his own views on atmospheric instability, and concluded by saying that he believed that he had the "instability problem by the tail." It was unfortunate that they were separated by such a great distance, because Rossby really wanted to talk it all over with Charney in person. He also had a directive for Charney: "condense the letter to von Neumann for publication in *Tellus*." By doing so, Charney would be stating the principal difficulties that they faced with the computer project, including a discussion of the significance of noise, high signal velocities, and the character of the barotropic model.[18] Rossby wanted to spread the word, and spread it quickly. Too little information had been coming out of the project, and he wanted to get these important theoretical developments in front of international geophysicists.

Sensing yet another opportunity to bring the project's work to the attention of the wider meteorological community, Rossby wrote again just two days later. This time, Rossby wanted Charney to write a brief note—based on the project's work on signal velocities—for the *Journal of Meteorology* to accompany a paper on energy dispersion by Rossby's "academic son," the Chinese meteorologist Tu-cheng Yeh (now known as Ye Duzheng). Rossby knew that Charney and his colleagues had had to determine under what circumstances a perturbation would be carried into the forecast region during the period of interest. To understand their concern, consider an extremely large (1,000 miles × 1,000 miles × 30,000 feet high) virtual "box" enclosing part of the atmosphere. To make a 24-hour forecast of the box's atmospheric properties, one would have to know how quickly atmospheric energy was moving in from the west (assuming mid-latitude west-to-east flow). If the inbound (horizontal) flow were moving at only 10 nautical miles per hour, then only those features within 240 nautical miles of the western edge of the cube would enter it. Features farther west could be ignored without adversely affecting the forecast. Therefore, in areas of extensive data coverage, there were sufficient data to make a one- or two-day prediction. Further, the vertical velocities were so slow that any stratospheric disturbances could not work their way down to the lower troposphere in this short forecast period, either. That meant available upper-air data were sufficient for the task—an extremely important consideration since additional (expensive) upper-air reports were not likely to become available just to satisfy numerical weather prediction requirements. Rossby thought that if Charney explained the significance of signal velocity to the computing project, he would educate readers to get away from explanations "in situ." Indeed, Rossby thought this was just the piece needed to enhance Yeh's work.[19] And, of course, just what was needed to alert meteorologists around the world to Charney's work in Princeton.

Rossby had immediately grasped that the obstacles the Meteorology Project was confronting—and the solutions they developed to overcome them—would be an important starting point in his drive to sell applied meteorologists on the importance of theory (and, equally, selling theoretical meteorologists on the importance of thinking physically). Advancing this agenda, Rossby once again offered Charney advice and direction, urging him to tackle the topic of internal waves that Charney himself had suggested in his letter of 15 September. (Charney, an infamously poor correspondent, had not yet responded to Rossby.) Rossby had attempted to work on internal waves in an incompressible atmosphere with either constant stability or a sharp density discontinuity. He had found stable waves and a small range of phase velocities. Again, he urged Charney to write an article for *Tellus* about the computing project. He did not want to overload the journal with theory and thought that a "clearly written exposé" about a computable model might help meteorologists gain a better attitude toward theory. He also asked Charney to discuss the difficulties of measuring approximations because it seemed to Rossby that "the majority of theoretical meteorologists hide their inability to think physically behind absurd insistence on 'accuracy.'" Rossby closed his letter by reiterating his belief that the Meteorology Project was important. But he was concerned about the "vast amount of housecleaning required in the storehouse of ideas among theoretical meteorologists and partly over the vastness of the educational task among the so-called practical meteorologists."[20]

Rossby wanted those studying with him in Stockholm to be thoroughly familiar with Charney's ideas and the project's progress. And he wanted to ensure that Charney knew that his European-based brethren were taking his ideas seriously. Yet his greatest concern was educating all meteorologists to the potential of numerical methods to advance both theory and forecasting.

Thus, Rossby had visiting Chinese meteorologist H. L. Kuo undertake a review of Charney's ideas on stability as presented to von Neumann. Penning yet another note to Charney, Rossby reported that the attendees had held an "extremely stimulating discussion" of his current work. Why, even those skeptical synopticians in attendance understood at last that they meant "business with the computing project." Again he tweaked Charney: get the letter to von Neumann cleaned up for publication—this time for the *Journal of Meteorology*—for the "education of meteorologists." Moving on to theoretical considerations, Rossby wanted theory to get back to fundamentals, i.e., theory needed to express the factors that forecasters use, either consciously or unconsciously, when making a forecast. Other terms, e.g., the divergence term, which either could not be considered, or were not considered, by the forecaster could be included at a later time, but for now should be eliminated from the equations. Assumptions that were already successfully used by forecasters, e.g., neglecting compressibility and non-adiabatic processes aloft, should be considered when formulating theory.[21] Rossby wanted to make sure that in their pursuit of theory, the Meteorology Project members—Charney in particular—took advantage of the knowledge already gained by those who dealt with the weather on a daily basis at the forecast desk.

Rossby had a wide network of contacts throughout the geophysical community with whom he maintained a regular dialogue. To keep his outgoing pipeline of information filled, he needed regular updates from Charney. What frustrated Rossby was that he was not getting them. In December 1948—having heard nothing from Charney in 3 months—he wrote again, chiding and nagging. Rossby was curious about the progress in Princeton. What was going on? He knew that Charney had not written the article for the *Journal of Meteorology* because editor George Platzman, a former Rossby protégé now at the University of Chicago, had neither heard from Charney nor gotten a manuscript in the mail. Rossby badgered Charney to get it out. And he pleaded once again for a summary for *Tellus*, not to compete with the *Journal of Meteorology,* but because the Swedish physicists and geophysicists needed to know that meteorologists were thinking in terms of calculating flow patterns.[22]

That Rossby was trying to get the word out to other scientific disciplines about the new, more theoretical approach in meteorology was apparently lost on Charney. When he got around to responding to Rossby with a long, newsy letter, he made it perfectly clear that he agreed with Rossby's philosophy of approaching meteorological problems, but was not going to write an article for *Tellus*. Charney assured him that he was indeed writing an article for *Journal of Meteorology* because it was better for "propaganda purposes." He saw no point in writing two articles about the same thing—an attitude that probably did not please Rossby.[23]

Operating on Rossby's research philosophy—once you think you have something figured out, try it on data to see what happens—Charney had the group consider an actual case. The starting point was a 500-millibar (mb) constant pressure map, i.e., a map which represents a surface in the atmosphere where the pressure is everywhere 500 mb—considered to be the halfway level between earth's surface and the top of the atmosphere. The lines on the map represent the height above the surface where the pressure is 500 mb. It varies from place to place, with the average height being about 18,000 feet (5,500 meters). (This differs from a surface weather map where the earth's surface is considered to be sea level everywhere and the equal air pressure lines—isobars—on the map create a pattern that varies across the surface. There, high numbers represent higher pressures, i.e., the weight of the air above that part of the surface is high. Low numbers represent lower pressures, i.e., the weight of the column is less than in higher-pressure areas.) The team members selected 500-mb-height values at 45° north latitude and inserted them into the formula derived by Charney and Eliassen.[24] They successfully predicted the deepening (pressure decrease) of a major trough (an elongated area of low pressure) in the central United States and the intensification (pressure increase) of a ridge (an elongated area of high pressure) in the eastern Atlantic. Since the technique was quite simple and had given such good results, Charney was convinced that it might prove immediately useful to forecasters, who would be able to predict a pressure profile for a specified latitude in less than 30 minutes. It was particularly important that the intensification and weakening of pressure features were explained solely as a result of the horizontal dispersion of energy. Although in some cases energy would come from above or below, it appeared that when considering the average motion, one could predict many features by using the barotropic assumption—a major simplification. However, one successful trial over the continental United States did not mean that the method could be generalized to other parts of the globe. So Charney had the team perform a similar trial run over the Pacific Ocean. Where the first trial was successful, the second was a complete flop. The initial situation showed an extremely long trough that should have been moved rapidly to the west. Indeed, their model forecasted it would do so. Unfortunately, to the surprise of the meteorologists (as sometimes happens in real weather situations), the trough actually stayed put. Project members were puzzled—not about why their model predicted movement that did not take place, but about why the trough did not move.

Although a barotropic model showed promise, Charney knew that it would not be the ultimate numerical weather prediction model. Why? Because the atmosphere is not usually barotropic. If it were, cyclonic (counter-clockwise turning, low pressure) systems and their associated fronts (boundaries between two distinct air masses) would fail to develop. It was when thermal and height patterns were not in perfect agreement—a state called baroclinicity—that storm development, to use a layperson's term, takes place. Therefore, Charney knew they would need to attack the significantly more complex baroclinic model. An intermediate model—an enhanced barotropic model such that the variation in the wind with height is averaged vertically—would be dubbed "equivalent barotropic." It was an ideal model to attempt because it needed only easily obtained height gradients as initial values. Charney had found that this model worked as long they used a 1-hour time step and 400-kilometer grid spacing. Once the team thoroughly investigated this model, Charney was confident that they could expand it to the baroclinic case. No matter what the investigation, Charney continued to follow Rossby's directions: try the simple version first and, once it works, increase the complexity. That is exactly what Charney intended to do.

Rossby continued his mission of closely following the progress of the Meteorology Project, providing advice whether it was asked for or not, and encouraging quick publication in *Tellus*—especially because he wanted to get the word out on the new scientific meteorology to all those physicists who doubted their scientific intentions. Writing in early January 1949, he expressed much interest in Charney's work on the extension of the energy propagation equation. However, Rossby was having a difficult time accepting Charney's conclusion that the barotropic convergence was of little or no importance. This was largely due to the "absurd result" in non-divergence theory that the western edge of a solitary disturbance is displaced with the speed of the zonal wind eastward that would be, in Rossby's view, much too fast. Disturbances just did not move at the same speed as the wind. And again, he reminded Charney that he was not only welcome to publish in *Tellus,* he was most strongly encouraged to do so. Rossby was extremely eager to show "pure" physicists that meteorologists were "getting out of fiddling" and developing significant theoretical approaches.[25]

Much to Rossby's consternation, Charney did not submit his "cleaned up" letter to von Neumann to the *Journal of Meteorology* until April 1949—a significant delay in getting the word out to the skeptical meteorological community. But undoubtedly hoping for better cooperation from Charney on his next paper, Rossby continued to provide venue and content publishing advice. While visiting Chicago, Rossby jotted Charney a note about the new

Charney-Eliassen paper. Rossby told Charney that von Neumann should write a preface for the paper giving a short explanation of the computer project itself and what it hoped to accomplish. The recommended publication venue: the *Journal of Meteorology*. In this case, Rossby was not pushing for publication in *Tellus,* because he thought it was more important to bring these new developments to the attention of the meteorological public in the United States than to publish in Sweden. He would, of course, print it immediately if Charney instead submitted to *Tellus*. Rossby further advised Charney to include samples of numerically predicted pressure profiles so that readers could see how the output looked.[26] In a follow-up note written while en route to Sweden (with the jotted postscript "please mark coffee as gift"— to avoid customs payment), Rossby suggested that the joint paper compare predicted and observed changes in the profiles because it would be a "severe test, but more fair" than comparing actual profiles.[27] He also telegrammed Charney, asking him to send a brief statement on what was happening at the Meteorology Project for the May issue of *Tellus* because it was "essential" to keep the progress of the computing project before the scientific public.[28]

By March, Charney had apparently put in a word with von Neumann about Rossby coming to Princeton, because the latter expressed his appreciation to Charney and reiterated that he would very much like to join the Meteorology Project. However, Rossby desired to maintain a lectureship in Chicago, which would allow him to keep close ties to its meteorology program and also work in Princeton with minimal travel back and forth to Chicago. He did not anticipate a large experimental facility being required in Princeton because he was counting on extensive cooperation with other institutions, particularly the University of Washington, Chicago, NYU, MIT, and UCLA.

Large-scale organizational and professional matters were much on Rossby's mind. He clearly wanted the opportunity to recruit a couple of younger meteorologists to do the needed synoptic investigations in order to continue the work in basic theory.[29] He thought it was "absurd" to set up an organization at Princeton that would compete with meteorology departments at universities. Instead, Rossby envisioned a totally cooperative relationship with both academic departments and government agencies.[30] He recognized that there were not enough academic meteorologists to go around, nor enough graduate students to fill their programs. Adding another formal meteorology department would just take away from the others for no net gain to the community of researchers.

In the interest of getting their results out sooner rather than later, Charney and Eliassen decided to submit their joint paper to *Tellus* instead of to the *Journal of Meteorology*. Unfortunately, printing problems delayed its appearance until June 1949. The Charney-Eliassen article, detailing the effectiveness of numerical weather prediction techniques for both theory development and possible weather prediction applications (the previously discussed forecast of 500-mb heights), was extremely important. Although it did not have an opening section written by von Neumann as Rossby had suggested, it was the first paper to give positive, concrete results from the Meteorology Project. Rossby thought that the article presented a new era in meteorology where " 'feeling' would be repressed in favor of computation." It also presented a more heuristic view because the team was willing to try an approach and see how it worked with actual data before determining the next move.[31] The theorists would be pleased that feeling was taking a back seat, at last, to more mathematical techniques. And the applied meteorologists would be glad to see that they were trying the newly developed theory on actual data for a reality check before moving on.

::

Rossby, being Rossby, had more pots to stir than just the Meteorology Project. In early May 1949, he invited a veritable who's who of European meteorology to Stockholm for a week of talks and discussions (and probably arguments) on climatic fluctuations and other related subjects—followed by the customary Rossby-led excursion to the Swedish countryside that accompanied all Rossby-inspired meetings.[32] However, never passing up a chance to preach the results of the Meteorology Project's numerical weather prediction work to the unbelievers (or at least the skeptical), Rossby took time to brief the assemblage on the Charney-Eliassen paper and its predicted pressure profile diagrams.

Considering the current state of the project and some concerns of his own and others that had been tossed around, Rossby wanted to know if it would be possible to make a numerical study of stationary wave patterns, which could very well inform climatological questions. If such a study were possible, did Charney intend to attack it? If not, Rossby had a couple of young meteorologists ready to work on it, but he did not want to duplicate research that was already spoken for by the Meteorology Project or the Weather Bureau.

When contemplating how best to put applied meteorology on a firmer, i.e., more scientific footing, Rossby had come to the conclusion that the Meteorology Project needed to "push the present approach" to its ultimate conclusion—an operational forecast—to swing the doubters into the numerical weather prediction camp. He reported that the British meteorologist Reginald C. Sutcliffe, for instance, had wanted to know why they could not simply extrapolate troughs and ridges from one day to the next based on past displacements, since they would come up with the same answer. To Rossby, the question was not even important. However, he thought it best if Charney knew the opposition—the very conservative people in the meteorology community who were going to be difficult to convince of the efficacy and desirability of numerical weather prediction. As a postscript, he recommended the Charney take a look at prominent Swedish oceanographer Vagn Walfrid Ekman's ocean current theory because it was similar to what Charney was trying to do more generally in the atmosphere.[33]

By late spring 1949, Rossby was spending most of his time in Stockholm. However, he continued to maintain close ties by mail with his Chicago colleagues—in particular *Journal of Meteorology* editor George Platzman. Writing to Platzman at the close of the climate change conference, Rossby wanted to sound out his friend on a number of ideas related to the Meteorology Project in general and the production of operational forecasts by numerical means in particular. He also knew that whatever he wrote to Platzman would ultimately get to Charney and thereby double the impact of his message. Rossby wrote Platzman that the Charney/Eliassen methodology—using a barotropic atmosphere with barotropic convergence and assuming a constant zonal current to develop a method of integration—seemed "extraordinarily promising." Their introduction of frictional forces in the model had prevented resonance difficulties. Although Charney and Eliassen argued that the method had practical applications because of the amazing agreement between observed and computed results, Rossby thought the method would break down when faced with rapidly deepening systems. However, he had shown the results to visiting Weather Bureau meteorologist Jerome Namias—Rossby's former co-worker and student at MIT. Namias had subsequently written to his brother-in-law, Harry Wexler, at the Weather Bureau, asking him to contact Charney and try the method in the forecast section. Although Rossby would have preferred a more rigorous test performed in an academic environment, he considered a Weather Bureau test to be better than no test at all.

Additionally, Rossby wanted to find a way to expand the work. He thought the calculations should be done for all middle latitudes (i.e., 35°, 45°, and 55° north), not just 45° north, and for several different values of zonal currents. If the values were then computed for different altitudes, they could be pieced together to create a three-dimensional view of the atmosphere. He also wondered if equal success would be reached by looking at moving versus stationary systems. If the method worked for moving systems, then there was a possibility of getting out of the "horrible subjectivity" that characterized "all or most" forecasting. Again, Rossby had pointed out that he had a couple of young men in Stockholm who could be employed on such a task, but claimed he did not want to "interfere" in US efforts. However, if all of the groups could work cooperatively and obtain results faster, that seemed to make the most sense.[34]

Leaders of research schools are constantly on the move—making sure that their acolytes' works are spread far and wide, and keeping up the flow of advice and moral support. They also push hard to see desired research and experiments achieved. Rossby's aggressive sharing of the Charney-Eliassen paper with the "younger people" (graduate students and post-doctoral researchers, presumably) and the Meteorological Institute's many visitors before its actual publication is an important example of this trait. Reassuring Charney that everyone who had seen it was very interested, Rossby was most anxious to get it on the street.[35] Charney genuinely appreciated this routinely received moral and technical support. He acknowledged his debt to Rossby's influence, writing that if he and Eliassen had been successful in the application of the heuristic method, it was because Rossby had taught them very well.

Rossby, recall, had also pushed the idea of an operational test in letters to both Charney and Platzman. By sharing the paper with Namias and suggesting the same operational test to him, Rossby was counting on Namias to make the same proposal to Wexler, and then execute the project once he returned to the Weather Bureau in late spring. Charney's hand was all but forced. Rossby knew that the chronically short-handed Weather Bureau would want to try out an objective technique that provided a prediction in half an hour's time. Once the possibility was out in the open, the Weather Bureau would be clamoring to try it no matter how reluctant Charney and Eliassen might be to subject their new method to an operational test. After all, if it did not work as advertised, it could set back their efforts to convince the applied meteorologists of the ultimate usefulness of numerical techniques. Furthermore, Charney did not want the practical applications to overshadow the important theoretical results: topography and friction were only minor players in short-range atmospheric variations. Thus, the models could ignore them in short-range predictions. Perhaps even more important, the results indicated that non-linear barotropic models would lead to both practical and theoretically valuable outcomes.[36]

The testing that Namias had proposed to Wexler was carried out at Princeton by Weather Bureau personnel. Charney and Eliassen went to bureau headquarters and delivered two lectures about their method. The response was so enthusiastic that Charney decided he needed to temper his remarks so as to not inadvertently oversell it. The testing team was to make pressure profile forecasts for several different weather types using the Charney-Eliassen equations for periods of two to seven days, and some for even longer periods. Charney noted that the Rossby formula yielded reasonable values for the displacements of the five-day mean pressure systems and therefore might be able to approximate dynamically possible flow patterns.[37] These tests would force the Princeton group to concentrate on longer period phenomena for which new physical factors would have to be taken into account.

Despite Rossby's desires, Charney thought there would be problems in extending their results. He and Eliassen had avoided some of them by assuming a constant, zonal basic flow, but other assumptions did not seem natural. Friction was unimportant for short period weather changes, but extremely important for stationary systems. Also, friction implied that system energy would be dissipated. However, the model could not dissipate energy unless it was provided with an energy source. The team decided to provide the energy by assuming the zonal current was maintained by a thermally driven meridional circulation. They also neglected any energy loss through perturbation flow.

Rossby had suggested that atmospheric baroclinicity could be introduced as an external factor in the two-dimensional model. Excited by this suggestion, Charney promised to keep it in mind. He was also pleased that the plans he had already made to investigate the stationary perturbation pattern were similar to Rossby's ideas on the subject. However, Charney had not planned to determine the stationary pattern for jet stream flow by superimposing the patterns for different parts of the stream, and was delighted to let Rossby's Stockholm group work on that.[38]

Since his arrival in Princeton, Charney had enjoyed the relative luxury of a stable personnel situation characterized by an ideal mix of disciplinary expertise. Project members had been simultaneously focused on the physical description of the atmosphere such that prognoses could be developed from available data, and the invention of a computing technique that would be both stable and responsive only to meteorologically significant atmospheric motions. Eliassen had been particularly helpful in both the physical and mathematical aspects of the project.[39] But Jacob Bjerknes had offered Eliassen a position at UCLA for the remaining months of his one year leave of absence from the Norwegian Meteorological Office, and Eliassen—desiring to take his bride to another part of the country—accepted. Charney acknowledged that while the UCLA opportunity was wonderful for Eliassen, he was going to miss him very much. The project would need a replacement with Eliassen's combination of theoretical and practical experience. Charney found this in Ragnar Fjørtoft of the Norwegian Meteorological Institute. Fjørtoft had worked with Charney during the latter's year in Norway, and Charney knew that he would be able to fit nicely into the project. Still looking abroad, Charney also invited British meteorologist Eric Eady to join the project. Having heard nothing by way of reply, he asked Rossby to check on Eady and find out what his plans were. And as for Rossby himself joining the Meteorology Project, no decision had yet been forthcoming from the Institute of Advanced Study.[40]

Rossby had played a major role in the Meteorology Project since its conception, but he had done so from off-site. Charney, longing for the days in Chicago when he and Rossby would spend many hours discussing meteorological problems, very much wanted Rossby to come to Princeton and become the project's director. Instead of waiting for letters to make their way back and forth across the Atlantic, they could sit down over strong Swedish-style coffee, discussing their ideas face-to-face. Rossby's close ties with Reichelderfer at the Weather Bureau, the academic meteorologists associated with the MIT and Chicago programs he had founded, and geophysicists from many disciplines throughout Europe, would aid in bringing in outside advice and support when they needed it. And so Charney had encouraged von Neumann to bring Rossby to Princeton.

After several months of negotiations within the Institute, von Neumann—with the concurrence of IAS Director J. Robert Oppenheimer—extended an invitation to Rossby to become a member of the Institute for 2 years. As von Neumann noted, Rossby "more than anyone else" was responsible for getting theoretical meteorology work started at the IAS under the auspices of the ONR contract. Although they had had a slow start, the pace had accelerated since Charney's arrival (and due to the "advice and encouragement" of Rossby). The computing machine was expected to be operational in early 1950. Von Neumann wrote to Rossby: "[Our] work will need your advice, and to the extent to which this is feasible, your presence, more than ever. In fact, we embarked upon it originally in the inarticulate but definite hope, that we should have your help and guidance, when we had developed the necessary tools, and come really to grips with the main problem." The proposed two-year contract would give Rossby sufficient time to carry out a "well-rounded" portion of the research program in theoretical meteorology and would give all of them enough time to come to some agreement about their "mutual possibilities and plans."[41] In other words, the Institute was not willing to bring Rossby on contractually for too long a period in case it did not work out. Given the very strong personalities that were involved in this project, von Neumann and Oppenheimer may have been reluctant to bring in yet another alpha male for an extended stay.

Charney, trying to convince Rossby to come to Princeton, worked to put the negotiations in a positive light. The permanent members of the Institute had had little or no knowledge of Rossby and his qualifications. Therefore, the process had been slowed down while von Neumann brought his work to their attention. Charney reported that after this "indoctrination" the decision to extend the offer was unanimous. Although von Neumann wanted the appointment to be permanent, it was thought best to leave the decision about the future open to both Rossby and the Institute leaders once they had become better acquainted. Von Neumann believed that "the going would be smoothed if an engagement period were allowed to precede the marriage." Rossby would become the head of the Meteorology Project, but would be able to maintain contact with the University of Chicago and other meteorological institutions. In his final pitch to persuade Rossby to come, Charney wrote, "You know as well as I that meteorologists will continue to be frustrated at every turn as long as they lack the mathematical ability to carry their physical arguments to their logical conclusions. I would like nothing better than to be able to help you to break this dam."[42] In the midst of this invitation, catastrophe struck the University of Chicago's Meteorology Department. Chairman Horace Byers had suffered a heart attack. Rossby, still tied to Chicago, volunteered to fill in for the summer before returning to Sweden to teach in the fall. Despite establishing a meteorology program in Stockholm, Rossby's intent in summer 1949 was to return to the United States permanently in early 1950.[43]

The Princeton group, anticipating the day when their new computer would be ready, had started to make tentative computations of 500 mb wind forecasts on desk calculators by early 1949. Meanwhile, down in Washington, the Weather Bureau's Joseph Smagorinsky was coordinating the future Princeton-based tests of the Charney-Eliassen numerical forecasting method. Since it was too difficult to do the needed calculations as the month unfolded—i.e., in close to real time—Smagorinsky and the Extended Forecast Section elected to use June 1949 data with a normal (i.e., climatic average) value for the June zonal current. Smagorinsky was convinced that the normal value would be unreliable because it would not be related to the June 1949 conditions. However, if they were calculating daily forecasts in June as the month unfolded, they could not very well use the June average, which would not be available until the end of the month. Smagorinsky thought they ought to try the calculations both ways: with the normal and with the calculated average value for the month, and then compare the results. If the normal values provided a solution that was close enough to that provided by the June 1949 average value, then in the future it would be possible to use the normal values under operational circumstances. The Weather Bureau's analysis center hoped to produce 36-hour prognoses eventually, but in the near-term would work on 24-hour prognoses instead. Due to time and manpower constraints, they had not gotten as far as the Extended Forecast group. Analysis center personnel would need to compile, plot, and analyze a tremendous amount of data before making the forecasts.[44]

ENIAC to the Rescue

By late summer 1949, the Meteorology Project's future progress depended on the availability of an electronic computer for trial runs of their models. Unfortunately, the IAS computer was not ready. So as not to lose time while waiting for the new computer, Weather Bureau Chief Reichelderfer intervened with Army Ordnance on behalf of the Meteorology Project.

Army Ordnance controlled the ENIAC—the special purpose electronic computer designed and built to solve ballistic problems. Unlike the new IAS machine, the ENIAC was not a fully stored-program machine. Programs had to be broken into small pieces and set on switches. Therefore, it would take a considerable amount of time to write a program specifically for ENIAC and to put the program into the machine. However, a slow electronic computer was still much faster than a hand calculator, so ENIAC was the best alternative available to the Meteorology Project.

In September 1949, Reichelderfer formally requested the use of ENIAC (located at the Aberdeen Proving Grounds, Maryland) for the Meteorology Project's barotropic model run. He supported his request by pointing out that the work being done at the IAS on numerical forecasting was of the utmost importance to both civilian and military interests. World War II had led to the establishment of many more surface and upper-air observation stations, which had greatly increased the amount of data available for weather forecasting. The additional data needed to be included into both meteorological analyses and prognoses. The electronic computer was, therefore, the best hope for helping to sort out all these data and solve the relevant equations that govern atmospheric behavior. Although the Office of Naval Research had initially been the only financial supporter, it had been joined by the Air Force. Military leaders had become increasingly aware that this was a project of strategic and tactical importance.

By using hand-calculators and human "computers," the Meteorology Project had already successfully predicted the 24-hour change in the 500-mb-height field when treated as a one-dimensional problem. Attempts to solve the two-dimensional problem by hand had been abandoned because it was too labor intensive. If the ENIAC were available, von Neumann estimated it could make a 24-hour forecast, calculated in 1-hour time steps, in 6–8 hours. To check forecast accuracy, they needed to make daily forecasts over a two-week period. Reichelderfer asked: could Army Ordnance make the ENIAC available for a two-week period sometime in the upcoming 3–4 months for the first application of electronic computing to weather forecasting?[45]

The Army responded swiftly. Noting "the importance of weather forecasting for military and civilian purposes," Army Ordnance granted permission to use the ENIAC for a two-week period on a not-to-interfere basis.[46] Reichelderfer forwarded this response to von Neumann, adding his "personal appreciation of the interest" shown by von Neumann in the "solution of the meteorological problem."[47] Von Neumann warmly welcomed this development, writing Reichelderfer that he felt "obliged" to him for his assistance in obtaining ENIAC and "this additional manifestation" of his interest in the work of the Meteorology Project. Von Neumann would be at Aberdeen, Maryland, for a meeting of the Scientific Advisory Committee of the Ballistic Research Laboratories in late October and would then make detailed arrangements for ENIAC.[48] Considering his close ties with the Office of Naval Research and the support he was receiving from the Air Force, it is curious that von Neumann sought Reichelderfer's—and not one of his military contacts'—help in securing ENIAC's use. Yet past accounts which assert that von Neumann rather than Reichelderfer made this crucial connection are in error.[49]

Meanwhile, in Stockholm, Rossby was trying to get some information from Platzman and Charney. He decided to take an "I'll fill you in, if you'll fill me in" approach when he wrote to both of them in October 1949. Rossby reported that visiting Belgian meteorologist Jacques Van Mieghem had been giving lectures on hydrodynamic instability. Almost all the others working at the Institute were addressing the formation and impact of blocking systems, i.e., high-pressure systems that remain in place and block the movement of atmospheric waves. US Navy Commander Daniel Rex—who 3 years earlier had arranged ONR funding for the Meteorology Project—was now working on his PhD in Stockholm under Rossby's supervision. His research focused on a comparison of blocking situations in Europe and North America. Having updated Charney, Rossby asked for a progress report on the project's nonlinear attack on the forecasting problem. He noted that his Finnish colleague Erik Palmén was still skeptical and very much concerned that numerical methods had too little connection with real atmospheric conditions. But Rossby thought it was healthy to win over skeptics, and he wanted more ammunition with which to do so.[50]

Smagorinsky continued his work on forecasting for the standard latitudes. Having overcome some initial difficulties, he anticipated having significant results before Charney's two-dimensional, non-linear model was ready.[51] Charney invited Smagorinsky to join the Princeton group in the three weeks preceding the ENIAC test so that he could become acquainted with the planning and coding. He suggested that Smagorinsky obtain a publication on coding so he would be somewhat familiar with the process before he arrived.[52] Smagorinsky was already trying to do this. He had also identified a possible scenario for the test run: the period starting 22 November 1949, when a massive block of high pressure suddenly appeared in the North Atlantic. The block persisted until 1 December, then weakened and diminished to the point that, by the time he penned his letter a week later, it had all but disappeared. Smagorinsky thought this two-week period would adequately test the two-dimensional finite-amplitude forecasting technique and would allow a good test to see how the barotropic model handled the blocking scenario.[53]

To become conversant with ENIAC, Charney had had to become, as he put it, a "servant" to the machine. Writing to Rossby on 21 December 1949, Charney opined that until such time as they had settled on one method and had thereby limited their options, it made the most sense for the person formulating the problem to be the one doing the programming and coding, or at a minimum, to maintain close supervision over the efforts. He also made this prophecy: "[In the future] the training of every meteorologist will include a course in numerical methods and the use of large-scale computing instruments."

Preparations for the ENIAC "expedition" required writing a computation scheme, and then translating that scheme into machine code. However, the scheme had to fit the machine, which had an internal memory of only 15 ten-digit numbers. Their model would require the storage of as many numbers as there were grid points, of which they had many more than 15. Therefore, they were going to have to use punch cards as external memory. A grateful Charney gave credit to von Neumann for his help in this regard.

Von Neumann and his wife, Klari, were also helping with the machine coding. For ENIAC, that meant one instruction for every ENIAC operation (in the order in which it occurred) had to be set on dials on the machine. The tentative plan was to go to Aberdeen and try it out in February 1950. Because Platzman had made major coding contributions, Charney hoped to entice him to come from Chicago to Princeton.[54]

The ENIAC plans were slowly coming together. Once again, however, personnel shortages threatened to disrupt the project. Several people Charney had thought were en route to Princeton had decided not to come. Charney was disappointed and surprised when von Neumann and Oppenheimer informed him that Rossby, disliking the trial "engagement," had declined the Institute's interim two-year offer.[55] Likewise, von Neumann was disappointed that Eady would not be joining the Meteorology Project in the summer as previously planned. He still hoped that Eady would join them once the IAS computer was operational.[56]

Changes to computer programs also contributed to delays in preparation. Von Neumann continued to handle the numerical analysis-related parts of the computer program, developing code to solve the model's equations. He found a method adaptable to the ENIAC, established the nature of boundary conditions, worked on stability criteria, and determined the influence of energy being propagated into the forecast area.[57] The ENIAC trials slipped into March. Charney, writing to Rossby, asked if he would be available to come to Aberdeen in mid March, after they had made one complete computation, so he could look over the results. Charney really wanted to see Rossby, and suggested that he might be able to get von Neumann to Aberdeen at the same time.[58] Rossby, then in Chicago, was eager to join them and tentatively scheduled himself to arrive on 13 March. However, he still reserved the right to change his plans at the last minute.[59]

::

Noon on Sunday, 5 March 1950, was the starting point of a 33-day experiment that became a major milestone in the history of the atmospheric sciences: the first computer-assisted attempt to forecast the weather by numerical means. The full-time expedition members (Charney, Fjørtoft, Freeman, Smagorinsky, and Platzman) ran three eight-hour shifts, five days per week, for five weeks (figure 5.1). The team, with the aid of ENIAC, produced two 12-hour and four 24-hour forecasts from initial observed data. As one participant later recalled, they encountered myriad difficulties with the ENIAC. On average, it could run error free for only a few hours and then took many hours to repair—with 20 accumulators each containing more than 500 vacuum tubes, there were many parts that could fail. The card-punch equipment was also prone to failure, although its mean-time-to-failure rate was not nearly as high as ENIAC's. Coding errors surfaced. The original two-week window stretched to five to allow time for additional runs.[60] Reichelderfer

Figure 5.1
The first ENIAC "expedition," April 4, 1950. Left to right: Harry Wexler, John von Neumann, M. H. Frankel, Jerome Namias, John Freeman, Ragnar Fjørtoft, Francis Reichelderfer, Jule Charney. (courtesy of MIT Museum)

ensured that the Aberdeen staff knew of his appreciation for "making this historic occasion possible."[61]

Once the expedition was over, Charney and Fjørtoft spent nearly 3 months analyzing the results.[62] By June 1950, Charney had something definite to report to Rossby. Forecast accuracy varied greatly from day to day. The model had produced terrible forecasts from data that should have played to model strengths. Unaccountably good predictions had been produced by model runs on data that should have created poor forecasts. The spatial grid size turned out to be too large. Much like a net with large holes that allow smaller fish to escape, cyclones slipped between grid points, leaving the model unable to identify their pressure patterns. The time increment was small, so the calculations were correct, but the computation time was too long.

Overall, Charney was sufficiently pleased with the results that he thought the barotropic model would be useful for qualitative explanations of atmospheric motion. However, the team would need to move to the more complex baroclinic model to produce the quantitative predictions needed by applied meteorologists. The IAS machine would have a 1,024-word internal memory—not sufficient for a baroclinic model unless they could arrange for external memory. Project members would try to run a partially advective model proposed by Fjørtoft (and similar to Sutcliffe's advective model) on the same data to see if it would handle cyclone development. They also planned to work on a primitive equation model and theoretical wave and vortex barotropic models, even though Charney was personally more interested in pursuing baroclinic models.

Despite these important advances in model development and understanding, personnel arrangements were again occupying Charney's time. Sharing Rossby's philosophy on personnel, Charney thought that it was a much better idea to invite those who understood the reasoning behind numerical weather prediction and were glad to be part of a cooperative solution than those only willing to till their "own furrow." In Charney's opinion, Eliassen and Fjørtoft were solidly in the first group, while Hunt and Queney were just as firmly in the latter. The initial members of the Meteorology Project had all been in the latter group—that was why initial progress had been so slow. Eliassen and Fjørtoft also had the advantage of possessing a broad knowledge of synoptic meteorology which had prevented the project from "degenerating into mathematical sterility"—a concern that had haunted Charney from the beginning. Freeman and Fjørtoft were both leaving in July 1950—Freeman for Chicago—but Fjørtoft would return to Princeton in September for only 4 months. No personnel additions had been planned for after that time, which, Charney noted, was "bad." Rossby had suggested that British theoretical meteorologist Thomas V. Davies might be a good addition, and Charney was sounding him out. Charney asked Rossby what he thought about bringing Namias in for a few months. Smagorinsky would be coming part-time to take Freeman's place, but they still needed two others at "the idea level." He asked Rossby to give them some advice and come himself if possible.[63]

Having considered the errors in the ENIAC results, Rossby was convinced of the importance of understanding the atmospheric processes that had been at work when the computations failed. Examining the maps with his protégés Swede Bert Bolin and US Navy officer Daniel Rex, all were amazed at how good the results were, given the model's simplicity.[64] Rossby proposed having Bolin conduct a synoptic study of the meteorological scenario because he was both an excellent analyst and a sufficiently well-trained theoretician to be able to come to a theoretically sound conclusion. Bolin was leaving for Chicago within a month and would be available to work with the Meteorology Project starting in early 1951. Rossby recommended Bolin because, like Eliassen and

Fjørtoft, he possessed both the synoptic background and the theoretical background that Charney needed. Desiring to get the results out quickly in the meteorological literature, Rossby strongly encouraged Charney and Fjørtoft to write a note for *Tellus* about the ENIAC calculations.[65]

Charney had also expected Platzman, who had played an important role in the ENIAC preparations and expedition, to move from Chicago to Princeton. Platzman's boss, Horace Byers, had been pressuring Platzman to remain in Chicago. Ultimately, Platzman decided to stay, and so informed a disappointed Charney. Platzman wanted to turn Chicago into a research center that would increase meteorology's standing as a science. "I feel," said Platzman, "that academic meteorology in this country is still suffering from the trade-school blues"—this despite efforts by the American Meteorological Society and its leaders, most of whom worked in the academic sector, to turn meteorology into a professional discipline accorded the same respect as engineering and the physical sciences. Platzman was hoping that with Dave Fultz—another Rossby protégé who had earned his PhD at Chicago (1947) and was known for his "dishpan experiments" that provided tangible evidence of how the jet stream moved in the atmosphere—he could bring new blood into the field and raise research standards. Platzman wanted his students to look at undertakings like the Meteorology Project and to be inspired to pursue theoretical research.[66] As it turned out, Platzman was able to join the Princeton group for the fall quarter, and Charney's morale received a boost with the opportunity to work closely with the Chicago group.[67]

The two-dimensional barotropic model had been fairly successful, so project members continued on their heuristic path, setting their sights on three-dimensional models and the possibility of using the primitive equations in numerical weather prediction. In mid-summer 1950, Charney was investigating the upper boundary conditions for a three-dimensional model. Fjørtoft was making a theoretical study of a simplified three-dimensional model and, with Charney, was studying the statistical-mechanical properties of two-dimensional incompressible flows. The IAS computer, which was supposed to be operational in mid 1948, was still not ready in mid 1950. Although the Meteorology Project members had no idea when the computer might be finished, as soon as it was, they planned to perform additional integrations of the barotropic equations with smaller grid spacing, and to begin programming in three dimensions. Most importantly, they would continue their efforts to formulate a theory for the physical nature of atmospheric motion.[68]

While the Princeton team continued its work, another team was setting up shop in Cambridge, Massachusetts. This one was under the direction of Philip Thompson, formerly of the Meteorology Project.

Research in a Parallel Atmosphere

As an active-duty officer in the Air Force, Philip Thompson knew he could not stay with the Meteorology Project indefinitely. By March 1948, after being on board for a little over a year, he was negotiating for his next assignment. Thompson wanted to stay in Princeton at least until the end of the year so he could see his efforts "blossom and be assured that [they] will later bear fruit." One possible alternative was to become the military attaché in Norway, where he would be able to use his position to obtain, directly and indirectly, valuable technical information about meteorology and other more general hydrodynamical topics. As an outgrowth of the Bjerknes dynasty in Norway, the Scandinavians continued to maintain a strong research program despite many of their students having moved to the United States during the war. In Norway, Thompson would be able to tap into this research network and pass information of interest back to the United States. However, his true desire was to return to the Meteorology Project.[69]

In August 1948, Captain Albert Trakowski of the Air Force's Geophysics Research Division at the Watson Laboratories in Red Bank, New Jersey, asked Thompson to recommend someone to lead the division's Meteorology Section. In response, Thompson managed to eliminate from contention everyone who was a meteorologist. Charney was out because "it's not his style" and he had no talent for administration. The same applied, in Thompson's view, to Victor Starr. Haurwitz knew the literature but was "old school" and lacked administrative ability and inclination. Indeed, Thompson could think of "no one *in the field of meteorology*" to whom he would entrust an organization dedicated to fundamental research in meteorology. He therefore suggested that Trakowski look for a geophysicist, especially one who had dealt with a hydrodynamical field like oceanography and had a "casual" interest in meteorology. Thompson wrote:

If I may make a couple of general remarks, meteorology seems to repel those sensitive souls who like mathematical or otherwise rigorously scientific treatment and draws a great many fools who, undaunted by the extreme difficulty of the problem, feel that metaphysical methods will yield results where scientific methods have not – whereas, several other fields of geophysics, better developed and more scientific (simply because the problems involved are less formidable), attract many able men.[70]

This was not the only time that Thompson would criticize the abilities of prominent meteorologists in an attempt to have them removed from consideration for administrative positions. In March 1949, when the Geophysics Research Directorate had been looking for a director for the new Cambridge

Labs, Thompson provided his opinion on Rossby's candidacy even after learning that the position was going to someone else. While claiming that he respected Rossby as a "man and scholar," he then went on to strenuously argue that Rossby would not be appropriate because he was "too narrow" (despite the breadth of his published works), that his ability to bring in new talent had three stages ("seduction . . . incest . . . degeneration") which effectively stifled any original work, and his lack of attention to detail was counterproductive. "Furthermore," Thompson continued, "for your sake as well as ours, may I suggest that you look at these things from the viewpoint of those who will judge you?" Thompson closed with this: "In short, it is my opinion that we should aggressively seek candidates potentially less dangerous to us, perhaps sacrificing a little of the vitality which we like in Rossby, and soon."[71] What should be concluded about Thompson and his aims? The meteorologists dispatched by Thompson as being unacceptable leaders of the Geophysics Research Directorate's meteorological section were among the most prominent, creative meteorologists of their day. Any of them would have been able to fully understand—and criticize—Thompson's research projects. Perhaps all of these distinguished meteorologists shared one characteristic: the potential to thwart the brilliant, yet insecure Thompson's ambitions.

Whatever Thompson's intentions or motivations when he wrote the aforementioned letters and similar ones, his apparent disdain for his fellow meteorologists is important. His attitude colored his efforts in pursuing numerical weather prediction as a strictly Air Force project over which he would have absolute control. And when the time came to take numerical weather prediction "operational," Thompson's drive for absolute control would lead to an explosive confrontation and would cause much heartburn among the others who had worked for years developing the theory and techniques with the Meteorology Project.

Having eliminated everyone else from consideration, Thompson was offered and accepted the position as Chief of the Atmospheric Analysis Laboratory (AAL) of the Air Force Cambridge Laboratories. Thompson wanted this move kept "fairly dark" so as to "reduce friction—and the heat engendered thereby—in a system whose viscosity is admittedly quite high." He discussed his planned move and new assignment with John von Neumann, who was supportive and wanted to maintain his association with Thompson.[72]

When Thompson advised Harry Wexler of his new assignment in late November, he disingenuously claimed to be completely surprised by the offer, but that under the circumstances it was the best career move for him. He was sorry to leave Princeton and "a group of meteorologists whom I esteem as highly as any in the world," but that he would only be "divorced" physically from the project. He planned for his new lab to complement the work being done in Princeton and looked forward to closer cooperation between the research services of the Weather Bureau and the Air Force's Geophysics Research Division.[73] The latter was a remarkable statement, insofar as Thompson was on record as considering the Weather Bureau to have been of little help to the Meteorology Project.[74]

Once Thompson became the AAL's director, the Air Force began to develop its own extensive plan for meteorological research. In the "Proposed Plan of Air Force Sponsored Research on Meteorology and Closely Allied Sciences," written by Thompson and issued in January 1949, the Air Force staked its claim to being the savior of modern meteorology. Thompson felt the Air Force needed to synthesize a unified theory of meteorology because previous efforts in geophysics research had been uncoordinated and results were lacking. Research coordination would improve and better results would be forthcoming with the Air Force leading a "frontal attack" on the weather-related scientific problems. Lest higher authority misconstrue this to be "pure science" unrelated to operational needs, the plan noted the "diverse operational requirements" of both the Air Weather Service and the other Air Force agencies concerned with atomic energy, electronics, and guided missiles. Thompson's plan was high-tech and high-cost. Specialized rockets for upper-level observations and photography topped the list. With too few men trained in the field, graduate fellowships were needed to entice students away from more glamorous physical sciences.[75] Clearly, the Air Force, and Thompson, had big plans for meteorological research.

Philip Thompson was thinking well beyond his own corner of the atmosphere to an overall military research policy. The Air Force was a large consumer of technical information, so Thompson thought it would be more economical if it conducted its own research. One drawback was that in order to be scientific it had to be reproducible by a team that had not conducted the original research. If the research were classified (due to data, technique, or results), another similarly skilled team had to have appropriate clearances. Research that (for reasons of security) was under military control would be in a separate category from research that was simply supported by the military. The Air Force would be required to carry out its own research plan as a matter of survival. To be successful, such a research organization would require the assignment of Air Force officers who were also scientists and researchers. They would bridge the divide between the military side, which controlled the funds and assigned the problems to be solved, and the civilian side, which would spend many years working on longer-term projects. Although these officers would not stay in the research arena (a career dead end) forever, their training, Thompson argued, would later make them valuable as intelligence officers and attachés.[76]

Thompson's AAL, while still striving to work on meteorological theory and numerical weather prediction models, was very much concerned with making contributions to short-range and long-range military objectives. The AAL wanted to be the Air Force equivalent of the Meteorology Project. Like the Princeton group, the Cambridge lab planned to operate under the guiding vision that a "unified theory of atmospheric motion would, of course, provide the perfect instrument for predicting weather." Unlike its Princeton counterpart, however, this laboratory had a full-fledged weather detachment (with probably 10–15 enlisted men and one weather officer assigned) operating within its confines for the purpose of handling and storing large amounts of data. The synoptic analysis section took care of its analysis and verification needs.[77] This gave the laboratory an advantage over the Meteorology Project, which had no in-house analysis capability and had to rely on the Weather Bureau's analysis section in Washington to provide that function. However, the lab did not have—nor would it have—a computer; thus, any models it developed would have to be solvable with desk calculators. In addition to Thompson's numerical prediction investigations, the AAL sponsored a variety of meteorological research projects within the lab. It also sponsored research projects in university meteorology departments (particular at UCLA and MIT), and it would eventually be a co-sponsor (with the Office of Naval Research) of the Meteorology Project.[78]

By early April 1949, Thompson's group was preparing to produce forecasts for the one-dimensional quasi-barotropic atmosphere. The plan was to attack a large-scale weather situation and compare the numerical results with those of a standard forecast.[79] By the end of the month, Thompson reported that the results were looking "rather promising."[80]

Little of Thompson's correspondence for the remainder of 1949 and into 1950 discusses his numerical work. What is most interesting is the *lack* of correspondence on the ENIAC expedition. For all his protestations of remaining involved with the Meteorology Project, Thompson was certainly not involved with the ENIAC runs, nor does he even bring the subject up in his correspondence with Harry Wexler. This is important, for previous histories of the Princeton Meteorology Project have told a distinctly different story. Historian Frederik Nebeker has written that Thompson traveled to Princeton every "two to three weeks" to keep in contact with the group. This information was based on an interview with Thompson conducted by historian William Aspray. Yet this differs from an account Thompson gave in a separate 1987 interview. In that account, Thompson said he went to

Princeton "fairly frequently" and that he and Daniel Rex "of ONR" were monitoring the Meteorology Project, since they were providing the money.[81] However, Rex was in Stockholm studying with Rossby during this period. Moreover, if Thompson spent that much time in Princeton, it is odd that those trips do not appear in any of the voluminous correspondence left by either Thompson or Charney (or, for that matter, von Neumann). Neither does Thompson mention this close liaison in either of his written accounts of the early days of numerical weather prediction efforts.[82] More likely, the later oral accounts of "frequent visits" were an effort by Thompson to align himself more closely with the by-then famous Princeton group.

By fall 1950, Thompson reported that the AAL had been doing a lot of work on the theory of large-scale motions and on verifying the corresponding prognostic equations. Thompson himself had incorporated some baroclinic effects into his large-scale motion theory and had found analytical solutions of the two-dimensional prognostic equations. He was writing up this theoretical development for publication in a Geophysics Research Directorate publication by early 1951. The lab, however, had moved on to advancing the work on two-dimensional models and was concentrating on extending one-dimensional theory.[83] Thompson had complained about the lack of "airing out" of ideas while he was with the Meteorology Project in Princeton.[84] Thus, it is interesting that he was publishing his new research in Air Force technical reports without benefit of peer review. Although some research papers produced in military laboratories contained findings that were subject to security concerns (which created problems in getting military releases to publish), Thompson's work did not fall under that category. Therefore, publishing his findings in an Air Force report seems to indicate that he was reluctant to have his work come under the formal scrutiny of his fellow meteorologists. His paper, "Notes on the Theory of Large-Scale Disturbances in Atmospheric Flow with Applications to Numerical Weather Prediction," was not published until July 1952.[85] At this point, despite his brave assertions, Thompson was on the margins. The main work in numerical weather prediction was underway in Princeton and Stockholm, and would remain there.

Onward and Upward

The two years that elapsed between Charney's arrival at the Meteorology Project and the first ENIAC expedition were busy ones for the international network supporting this theoretical effort. Previous narratives of this project have focused almost entirely on the computer aspects. In fact, it was the meteorologists who controlled and advanced it. Meteorologists from academe, the Weather Bureau, and the military—in Princeton, Stockholm, Chicago, New York, Cambridge, and Washington—were busy trying to develop a workable theory of atmospheric motion that could then be programmed into von Neumann's new machine. While von Neumann's efforts related to computational methods and hardware development should not be underplayed, neither could he have successfully completed this effort without the leadership of Princeton-based Charney and the off-site but intellectually and spiritually present Rossby. The continuous presence of Scandinavian Tag Team members Eliassen, Fjørtoft, and (at the end of this period) Bolin provided the genuine sense of the atmosphere necessary to keep the project from being separated from the physical world and tumbling into a mathematical fantasy land.

In stark contrast, the Atmospheric Analysis Laboratory's numerical weather prediction section, headed by Thompson, basically had just Thompson himself. Without the direct path to publications such as *Tellus* enjoyed by his Princeton counterparts, Thompson's influence on numerical weather prediction was limited to the Air Force. A minor player in Rossby's research school, he had very little intellectual impact on the Meteorology Project after his late 1948 departure.

In the next period, both groups would seek to develop even better models as they moved toward operational numerical weather prediction. Not to be outdone by the Princeton team, another group was taking shape in Europe: Rossby's own International Meteorological Institute in Stockholm.

6 :: Creating a Realistic Atmosphere (1950–1952)

The success of the first ENIAC "expedition" gave the Meteorology Project a much-needed boost. With their simplified barotropic model providing output that at least *looked* meteorological, it was time for the team to *increase model complexity*. In an elaborate system of guess and check, they would introduce new techniques and variables, perform a test, compare the output to an analyzed weather map, and then verify the model.

The meteorology team would continue to develop more sophisticated models while the computer was being completed, and von Neumann and his Computer Project team would assist with mathematical solutions. The question remained: How would they know if the model had produced realistic output? While they were modeling, others would have to be generating a subjective (hand-drawn) product against which they would compare the model's output. No two subjectively created meteorological maps were (or are now) *ever* exactly the same. If they were going to decide on atmospheric reality by comparing their objective, computer-made chart against a subjective, man-made one, how would they know that either of them was right? Which reality would they choose?

The Meteorology Project at the Institute for Advanced Study was not the only group attacking the problem of numerical weather prediction. Carl-Gustav Rossby's Stockholm team was providing analysis support to the Princeton group, training US meteorologists (military, Weather Bureau, and academic) as well as those from Europe, and preparing its own eventual launch into numerical weather prediction with assistance from the Swedish Air Force. Philip Thompson's group at the Air Force's Geophysical Research Directorate in Cambridge, Massachusetts, was hard at work on his models. Besides pushing development of his own model, Thompson encouraged the expansion of Air Force meteorology research funding to the European theater as well as to the Meteorology Project.

Though there appeared, at first at least, to be no overt competition—no overt "race to the finish" among these groups—those who reach the goal first in the scientific arena usually take all the credit. For numerical weather prediction, there would be not only a theoretical goal of describing how the atmosphere worked, but also an *operational* goal of using the computer to create usable forecasts for military and Weather Bureau meteorologists. Would the interagency and international cooperation that had been a feature of the developmental phase extend to the operational phase, or would a single winner attempt to corner the market on providing numerical forecasting tools to the nation? Who would get to make the choice?

Out of the Blocks

The Princeton team members were not the only ones encouraged by the results of the ENIAC expedition. In August 1950, Francis Reichelderfer asked Harry Wexler about the Meteorology Project's progress since the team had returned home from Aberdeen. When would numerical weather-prediction techniques reach the stage at which "we should start a full operating unit [at Weather Bureau headquarters] to make use of the results of the Princeton research," he wrote, so that the bureau could request sufficient funds to cover the cost in that year's budget?[1] Reichelderfer was overly optimistic. A few more years would pass before a computer would provide operational output useful to bureau forecasters.

Indeed, the Princeton team took several months to analyze the ENIAC results. George Platzman (in Chicago) was in close contact with Jule Charney about the model he was developing with Ragnar Fjørtoft. While providing extensive feedback on the finer points of the model, Platzman became concerned that the addition of more and more physical influences into the quasi-geostrophic model was leading to a very complex computation—more complex than an all-out attack on the primitive equations might be. (In the quasi-geostrophic model, the time scale on which a system evolves is slow compared to the rotational period of the earth, and the length scale is larger than the distance cold pools of air can spread under the influence of the Coriolis force.) Platzman wondered if they ought to just abandon the other models and move directly to the primitive equations.[2] Of course the problems with the primitive equations were all too clear, instability being the greatest.

The ENIAC results were important for future model development, and Charney was eager to get them published and circulating within the wider geophysics community. By then the Meteorology Project had been working on numerical weather prediction for 4 years, and the ENIAC results presented the first concrete examples of computer-produced pressure surface forecasts. That was a significant achievement in its own right, but, even more importantly, it would give those actively promoting numerical weather prediction more ammunition against the skeptics—both theoretical and practical meteorologists.

The need for additional ammunition was becoming crucial because Charney's 1949 paper "On a Physical Basis for Numerical Prediction of Large-Scale Motions in the Atmosphere," which had laid out the Meteorology Project's theoretical underpinnings, had come under attack. Richard S. Scorer, a meteorologist at Imperial College, London, had written a letter to the *Journal of Meteorology* questioning a number of points, in particular the adequacy of the barotropic model, in Charney's paper.[3] Charney had to craft a response that would diffuse the influence of Scorer's comments on the assumptions underlying numerical weather prediction. He asked for Rossby's reaction to Scorer's criticisms and to his draft response. Charney also asked whether Rossby could publish the new Charney-Fjørtoft-von Neumann paper in *Tellus*'s November issue.[4]

Rossby swiftly assured Charney that the article would make the cutoff date if he would submit it in time. The Scorer letter was more troubling because it contained, in Rossby's view, numerous unproven assertions that Scorer had used to support his contention that the barotropic model should be abandoned. Rossby was particularly irritated by Scorer's argument that meteorologists should leave numerical methods behind because the models did not contain every atmospheric detail, including the effects of gravity waves—which, Charney had pointed out, complicate the computations without improving the solution. Rossby perceived the influence of Reginald C. Sutcliffe—a British meteorologist who opposed numerical weather prediction—in Scorer's attack. Of primary importance, Scorer had not taken into account the difference between linearized and non-linearized wave theories. The former, often used successfully to solve problems involving sound, gravity, and frontal waves, could still produce sharp wave fronts that were maintained much longer than one would observe in the atmosphere—a bad characteristic for a model to have. Rossby had found Charney's response to be "all too soft . . . for a man who is apparently being used by [David] Brunt as a hatchet man." Brunt (like Sutcliffe, a dynamicist at Imperial College) opposed numerical weather prediction. In particular, Rossby was upset by Charney's "uncalled-for concession" in declaring that it was a "happy coincidence" that the barotropic model approximated the atmosphere sufficiently to be of practical use in forecasting. Rossby found that the approximation was not a happy coincidence at all, but instead a dynamic requirement based on the physical nature of the atmosphere.[5]

Charney saw things differently. His greatest fear was that those associated with the Meteorology Project, Rossby included, would oversell the practical forecasting aspects of the barotropic model in particular, and numerical weather prediction in general. This, he explained, was why his response had emphasized the word "practical" in relation to forecasting—numerical output did not yet measure up to the skills of a solid forecaster. Charney did not want to give critics like Scorer any more ammunition in their war against numerical weather prediction, for he agreed that the barotropic model was important in theory development. Charney's original response to Scorer, by his own account, had been "scorching," and he had toned it down considerably so as not to appear defensive.[6] Rossby and Charney both recognized that they would have to consider questions of timing and presentation to the wider meteorological community if numerical weather prediction were to become credible to the majority of discipline practitioners accustomed to operating in a much more subjective manner.

Indeed, Charney soon reached a very wide audience of meteorologists and weather enthusiasts in the United States. This occurred when his paper on the progress of dynamic meteorology research, which he had presented at the January 1950 AMS Annual Meeting, was published in the September AMS *Bulletin*. Charney pointed out that numerical weather techniques were not likely to lead to a "revolutionary increase in forecasting accuracy." However, he did anticipate that numerical techniques could yield more accurate forecasts of large-scale patterns in the upper atmosphere. The steering flows derived from these patterns could then be used to make forecasts of smaller-scale frontal systems that produce weather. While it was certainly possible that numerical techniques could someday be used to predict cloud formation and resultant precipitation, extant techniques could not deal with that level of complexity.[7] Although the audience for his talk had been relatively small, the print version quickly reached several thousand people interested in the status of numerical weather prediction. Charney's carefully crafted remarks were planned to inform without overselling capabilities. This was important to his, and Rossby's, overall strategy for winning skeptics over to the new meteorology.

By late October, Charney was working to finish his *Tellus* article before the 1 November 1950 deadline. After checking with von Neumann (then in Los Alamos) to clarify the derivation of the computational stability criterion, Charney advised him that they had completed some calculations by using the three-dimensional model on a scenario that had produced a large error in the barotropic forecast. These new calculations explained the error, so Charney and Fjørtoft planned to include the results in the article. Since they had finished coding the barotropic model for the new Princeton machine, Charney wrote that he was ready to receive his promised reward. In particular, he was waiting for "that dinner that Klari" (Mrs. von Neumann) had promised him. However, things must have been just a little hectic in Princeton. Charney admitted he would probably need Klari's meal in "liquid form" by the time they returned.[8] Indeed, at the end of December, Charney and Fjørtoft were still working on the *Tellus* manuscript and hoping to get it into the next issue.[9]

In fall 1950, Wexler paid another visit to the Princeton team. He reported to Reichelderfer that they had tested the barotropic model and were now moving on to the baroclinic version. The new computer was "almost ready" to go, but no one really knew when it would be ready for testing—perhaps by year's end.[10]

Other agencies holding a stake in the Meteorology Project were also inquiring about the computer's status as the year came to a close. With the project's ONR contract set to expire 31 May 1951, ONR staff member and Navy aerologist Lieutenant Max Eaton needed to discuss an extension with von Neumann and Charney. However, it appeared that the terms might change—Philip Thompson had contacted Eaton about adding Air Force money to the pot.[11] Before the Office of Naval Research accepted the Air Force's cost-sharing offer, Eaton wanted to discuss it with von Neumann.

As the new year began, Rossby planned to spend nearly 2 months consulting with Meteorology Project members. His presence would be particularly important given the decisions that needed to be made on future model development. Project members expected that the IAS computer would be operational soon, and they wanted Rossby's advice on how to best make the transition to the new machine. In advance of his arrival, Rossby had written to Charney and expressed his hope that the long-awaited corrected manuscript on the ENIAC results was en route to Sweden. He was hoping that von Neumann's contribution to the ENIAC article would draw more mathematicians to geophysical research.[12]

Wherever Rossby visited, his charismatic personality invariably drew in members of his far-flung research school like a magnet. Therefore, his presence in Princeton provided the perfect opportunity for his former doctoral student, Harry Wexler, to pay a visit. Wexler not only wanted to pay a call on Rossby, he also wanted to assess the project's progress since his fall visit. Model preparation invariably depended on analysis assistance from the bureau. Wexler needed to determine what the team members would need and when they would need it. He reported back to Reichelderfer that project members had decided to forecast, as much as possible, in real-time based on numerically determined 2–3-hour pressure tendencies aloft, i.e., over time, whether pressures were rising or falling. Therefore, they would need twice daily standard pressure maps from the Weather Bureau-Air Force-Navy (WBAN) analysis center—a joint activity manned by all three weather services for the production of analyses and prognoses. The WBAN would mail the charts daily for the duration of the experiments, through the end of February 1951. The bureau would also provide a copy of the 700-millibar (10,000 feet, 3,000 meters) prognoses for comparison.[13] These maps would supplement those already provided by the bureau. Charney was extremely grateful to the WBAN for providing plotting and analysis services for the project, which had insufficient manpower to handle those tasks.[14] The bureau never had money to support the project, but even when short of manpower, its leadership worked hard to scrape together enough man-hours to get these necessary tasks done.

Wexler was not Rossby's only visitor. Leaders of the Air Weather Service's Research and Development branch were interested in the project, and wanted to provide funds. Specifically, the AWS wanted to know if Rossby were planning to remain in Princeton, and if the project would expand. Rossby was tied up with his new Meteorological Institute in Stockholm, so von Neumann and Rossby both envisioned only a modest expansion of the project over the next 6–12 months. What they really desired was "quasi-permanency" for Charney. The AWS quickly responded that its Research and Development branch could offer the Meteorology Project a five-year contract, renewable annually so the project would always operate under a sustained period of guaranteed funding. Further conversations between the AWS's Colonel Benjamin G. Holzman, Rossby, and Charney resulted in an offer to fund any expansion of the project, including sub-contracts to Rossby's Stockholm group. Rossby thought that such funding for Stockholm by the U.S. government would be fully justified because both Rossby and Eliassen could "profitably" spend several months of each year in Princeton. Furthermore, European meteorologists who were possible additions to the Princeton team could be hired by Rossby in Stockholm and evaluated on their potential usefulness before sending them on to Princeton for a year or two.

Developments were progressing so rapidly that not all interested parties understood the new arrangements. Von Neumann was not at all sure if ONR Director Alan T. Waterman had been briefed on increased Air Force interest, or the possibility of a sub-contract for Rossby. Charney and von Neumann had already briefed the Meteorological Panel of the Research and Development Board on 1 February 1951. The panel concurred in the concept that the Meteorology Project should receive additional support "within reason" whenever it was needed. However, it had not known of the possible five-year contract or extension to Stockholm.[15] Their interest in finding additional funding for the project might have been lessened had they known about the Air Force's offer.

In mid February, Wexler returned to Princeton—this time accompanied by Jerome Namias of the Extended Forecast Section and J. R. Fulks of the WBAN Analysis Center. Colonel Benjamin Holzman joined them. They wanted to see how Charney's group was using the WBAN prepared maps and how those techniques might be applied by Weather Bureau forecasters before the implementation of computer operations. The Princeton group used the charts to compute (by hand) the instantaneous height tendency field at 500 mb (18,000 feet, 5,500 meters) for the barotropic model and a special-case baroclinic model. The group members then created a 24-hour prognosis from the tendency fields to compare with observed data. To their delight, situations with distinct, large-amplitude troughs and ridges showed a good match between predicted and observed values. The new computer would significantly reduce the computation time, which was then 4–6 hours. Until then, Charney suggested that the WBAN perform the same procedure as an aid to their analysis and prognosis routine. But, Wexler noted, they would need to find additional manpower before it would be possible. Fulks promised to look into it.[16]

Here again, an old obstacle re-emerged: Fulks determined that, since the personnel situation was at an "irreducible minimum," the WBAN analysis center would have to find more people.[17] Upon further consideration, Fulks recommended that the Weather Bureau supply one additional person to work on the tendency computations and ask the military services to each provide one additional enlisted man. To this end, he suggested submitting a proposal to the Joint Meteorological Committee (Joint Chiefs of Staff) subcommittee that coordinated meteorological services.[18] However, when this request came up for review, the Air Force and the Navy declined to participate because the WBAN was not supposed to be used for "research projects." Therefore, the Weather Bureau assigned an employee from outside the WBAN to perform these calculations.[19] In fact, the tendency computations were not exactly a "research project" and the decision to make them came from within the WBAN, and not from an outside organization. So while both the Air Force and Navy were sponsoring the Meteorology Project, decision makers somewhere in those organizations considered a technique under evaluation to be research and hence unworthy of additional personnel support. The research and operational arms in the military services were probably not communicating effectively.

In late March 1951, the Princeton group submitted a progress report encompassing the period since 1 July 1950. It did not include the ENIAC expedition results, since Charney, Fjørtoft, and von Neumann had published them in *Tellus*.[20] However, the project members did report that the computations had been sufficiently good to confirm the usefulness of numerical methods for weather forecasting, although grid coarseness had most probably obscured model errors. They intended to rerun the computations with a finer mesh grid on the new Princeton computer. The new machine would require a modified program and recalculation of the stability criteria—von Neumann had accomplished both tasks. Moreover, George Platzman and Weather Bureau statistician Margaret Smagorinsky (wife of Joseph Smagorinsky) had completed the coding in December and were ready to run forecasts from actual and idealized situations as soon the computer was available. The ENIAC runs had demonstrated that conventional map analysis and subjective interpolation of grid values were not good enough for numerical computation. Therefore, George Platzman and Margaret Smagorinsky would also work on an objective analysis technique and the accompanying programming for the new machine.

The group had also spent a considerable amount of time "re-tooling" for three-dimensional forecasts, trying a variety of approaches that could potentially yield better results. While the two-dimensional models were helpful for understanding the atmosphere's physical nature, they had limited forecasting applicability. However, like all experimental models, these early versions of the three-dimensional models soon proved inadequate as well, and so project members looked ahead to creating a solution of the general three-dimensional equations. Von Neumann and Charney were actively investigating the best techniques for handling them.[21]

Even without the new computer, the project had continued to expand its efforts beyond its first simple barotropic model. However, team members would not be able to advance without additional computer runs.

ENIAC to the Rescue—Again

In April 1951, the Meteorology Project members were ready to run their latest models, but the IAS computer was still not operational. Once again, Weather Bureau leaders made the initial overtures to Army Ordnance to arrange for computer time and quickly passed the affirmative response on to John von Neumann.[22] But while the Army had not charged for computer time during the first ENIAC expedition, this time the project would have to pay $800 for each 24-hour day of use because the Bureau of Ordnance's budget no longer had sufficient funds to cover runs for outside agencies.[23] Von Neumann worked on getting the necessary additional funds from the ONR.[24]

The time it took to arrange a transfer of funds from the ONR to the Bureau of Ordnance further delayed the arrangements.[25] Yet if von Neumann and the project members were frustrated by this delay, they did not so indicate in their correspondence.

Justifying the project's need for time on the ENIAC, von Neumann advised the ONR that Charney and Rossby were jointly planning the second set of ENIAC runs. Not only would they be testing model effectiveness, Charney and Rossby would be investigating a number of theoretical ideas that were related to research questions concerning the zonal westerlies and the development of blocks. Studying idealized flows, they wanted to find out the kinds of interactions that would occur between finite-amplitude waves and mean zonal flow. Charney and Rossby also wanted to explore how kinetic energy from a solitary wave dispersed, depending on whether it was connected to an unstable or a stable zonal current. They wanted to know the circumstances under which an atmospheric jet changed from one that widens gradually to one that widens rapidly.[26] This second ENIAC expedition would allow Charney and Rossby to extend their ideas about the usefulness of numerical methods to theoretical as well as practical meteorology.

The ENIAC run was scheduled for the end of May. Project members had not made any final decisions on which models they would test, but they had determined that their overarching purpose would be to study the non-linear interactions of systems in hopes of gaining a better understanding of observed atmospheric phenomena. Charney posed a series of questions about the behavior of perturbations in zonal flow as related to perturbation wavelength. Although the group agreed with Rossby and H. L. Kuo that they should try to discover properties of the general circulation that could be explained barotropically, they were convinced that the models they had been using were too restricted and possibly physically unsound.

No matter what they decided to run, available computer time would be limited. Charney hoped for a minimum of 14 and a maximum of 24 integrations, given their allotted three-week time slot and assuming 90 percent machine efficiency. Considering the machine problems they had encountered during the first expedition, there was no guarantee that this series of computer runs would be any smoother, and hence the number of completed runs might not be as great as they had hoped.

Army Ordnance had promised them the ENIAC starting 28 May, but it soon appeared team members could not have it until after 30 May. That meant they would be working at the proving ground until after 15 June. Neither Jule Charney nor Bert Bolin could stay that long, so Charney suggested that George Platzman, former team member John Freeman, and

Platzman's doctoral student Norman Phillips all come down to Aberdeen from Chicago for the runs.[27] Since too many people might lead to confusion, Charney proposed that they rotate in and out of Aberdeen from Princeton. Those left in Princeton would work on analyzing the results or revising the models as required. Norma Gilbarg, one of the project's "computers," would perform hand calculations in Aberdeen. Joseph Smagorinsky and Thomas Davies would also be available to work in Aberdeen.[28]

The delay ultimately caused considerable scheduling conflicts. After discussing the matter with Charney, Platzman and Phillips developed a plan: they would leave Chicago for Princeton to arrive on 21 May. After completing some preliminary work, they would return to Chicago, and then head east to arrive in Aberdeen on 4 June. The participants' personal calendars became stumbling blocks when coordinating the next round of model runs. Everyone had someplace else to be: Freeman was scheduled to give a paper in Los Angeles, Phillips had to return to Chicago for his convocation, and Platzman was obligated to attend a wedding. Platzman thought he could manage the last week in Aberdeen with Joseph Smagorinsky, Davies, and Gilbarg.[29] Von Neumann would be in Los Alamos during this second run.[30] Despite these difficulties, the second expedition moved forward.

Results proved mixed. In mid July, Charney (visiting the University of Chicago) advised von Neumann (still in Los Alamos) of the second expedition's findings. ENIAC and IBM equipment failures adversely affected their operations during their last week in Aberdeen. In four weeks, they had computed 70 time steps for four different models. Charney estimated a 41 percent operational efficiency—considerably lower than the estimated 90 percent. Machine failures cost 40 percent of their time; programming errors another 19 percent. Nevertheless, there was positive news. The model run had contained wave perturbations on a west-to-east zonal current featuring a jet-like structure. Perturbation energy was fed into the mean current for two of the models and withdrawn from the other two, giving results that qualitatively matched their preliminary calculations. Charney and Smagorinsky worked up the theoretical analysis of the results, and Bolin wrote up the barotropic tendency calculations. By July 1951, Charney and Bolin were looking for an appropriate journal in which to publish their results.[31]

::

After earlier toying with the idea of remaining in the United States after a fall of teaching in Stockholm, Rossby had decided to stay in Sweden permanently, determined to establish an International Meteorological Institute.

Initially, in the early 1950s, Rossby's institute was a small affair. Rossby, ever dreaming grandly, had envisioned a larger organization; however, since he would be setting it up "without an angel [a major patron]," he settled on starting with a smaller group. This proved a benefit to the Princeton group: able to support fewer researchers at the institute than he had anticipated, Rossby could spare some of his people to work elsewhere. But lack of sufficient funds also meant that Charney might not receive anticipated assistance from the Stockholm group.

As an interim measure, Charney proposed that von Neumann appoint Arnt Eliassen—who had been the first member of the Scandinavian Tag Team at the Meteorology Project—to the IAS staff on a part-time basis. Then Eliassen could shuttle between Princeton and Stockholm while being paid full-time by the IAS. Charney needed Eliassen. "Half an Eliassen," he told von Neumann, "is better than two non-Eliassens, and far better than no Eliassen."[32]

Charney was not sure about Rossby's plans, other than that he intended to teach in Chicago during spring term 1952. He also knew that Rossby had been trying to secure endowment funding from the Munitalp Foundation, a New York-based organization supporting basic meteorological research.[33] However, that possibility was fading rapidly. If Rossby's group failed to obtain Munitalp money, Charney suggested that they might want to apply for the funds on behalf of the Meteorology Project. Rossby, always on the lookout for meteorological research funding, liked this idea, and expressed his willingness to intercede on their behalf.[34]

The project's ability to evaluate the impact of the jet stream got a huge boost in summer 1951. It was then that Joseph Smagorinsky succeeded in finding a way to solve the partial differential equation governing the influence of continental topography and friction on a jet-like westerly current. His discovery meant the team now would be able to evaluate the effects produced by displacement and intensity changes in the zonal jet stream. The Princeton team also thought topography-influenced perturbations would localize centers of action in the upper atmosphere. This would provide a way to introduce heat sources and sinks into their models, as well as large-scale turbulent stresses responsible for variations in the mean jet stream structure.

Meanwhile, Charney had been working on a theoretical investigation of the mechanism which transfers angular momentum and kinetic energy in both barotropic and baroclinic wave systems. He found that both stable and unstable baroclinic perturbations increased zonal kinetic energy. Charney had examined four barotropic cases. In each one, he had found a net increase in momentum in the middle latitudes.

The British Meteorological Office (BMO) had also been looking at the quasi-geostrophic advective model and had found it wanting—as had the Princeton group. However, when they included the vertical advection of entropy in the calculations, the results agreed with observations. BMO personnel had shared these discoveries with Charney during his late-summer-of-1951 tour of European institutions conducting research on numerical weather prediction, and Charney had heard similar comments from other groups. Therefore, the Princeton group concluded that it would take the integration of the complete three-dimensional quasi-geostrophic model to produce a major improvement in forecasting. They decided to turn their efforts toward programming this model for the Princeton machine. Rossby's Stockholm group would do much of the preliminary testing and the synoptic analysis.[35]

Charney continued to be concerned about the "fog of criticism" that had begun to settle around their numerical efforts, and firmly believed that they needed to lift it very quickly. Writing to Rossby in fall 1951, Charney opined that the heart of the criticism stemmed from the Meteorology Project's reluctance to embrace the baroclinic model. Charney thought that if they stayed with the so-called two-dimensional or 2½-dimensional baroclinic models they would end up with the same inconclusive results they had seen from the barotropic models. Granting that there was still much to learn from these models, he greatly feared a loss of priority to other numerical prediction groups who were just then starting to work on them. If the Princeton group did not move into three-dimensional modeling soon, they could find themselves overtaken by other groups.

As Charney began to think about a question von Neumann had raised about the number of different weather parameters they were considering as data, he had an epiphany: the team was not including that many different weather elements in their model, so the new Princeton computer would be able to handle the three-dimensional problem. At that point, project members decided to start immediately on programming the three-dimensional baroclinic model for the new machine. They soon discovered that potential temperature was a better coordinate than pressure—which Rossby had told Charney some time before, but the latter had "pooh-poohed" the idea. Charney now asked Joseph Smagorinsky to find the appropriate equation. They found potential temperature worked better conceptually and computationally than using pressure or height as the dependent variable. Then they found out that Weather Bureau meteorologist Frederick G. Shuman had done his thesis on a related topic. Wexler sent Shuman to visit the Princeton group, but persistent personnel shortages at the Weather Bureau complicated matters; Shuman was unable to follow up on his ideas because he had been assigned to be a tornado forecaster.[36]

Although it might appear that the team's programming difficulties were mostly technical, some physical concerns remained. One of those dealt with how to treat the tropopause—the isothermal layer between the top of the troposphere and the bottom of the stratosphere. If they treated this layer as a discontinuity, handling the resulting boundary conditions could be a significant challenge. If the team considered the tropopause to be isentropic (i.e., of constant potential temperature), the boundary conditions would be easier to address. However, the tropopause was not always isentropic. After considering their options, team members decided to assume that it was not a discontinuous layer.

The project's major predicament was having too few people to perform too many tasks. With Joseph Smagorinsky spending all his waking hours on his dissertation, Charney and Phillips were doing most of the work. They desperately needed synoptic help, but instead found themselves with more mathematicians. The Bureau of Standards had loaned them a mathematician who had been a huge help on some mathematical problems involving the inversion of matrices. He had asked to join the group, but Charney thought bringing in yet another mathematician was akin to "carrying coals to Newcastle." Eady had proposed sending his student Andrew Gilchrist from the United Kingdom, an offer that was quickly accepted. Norman Phillips was working out "splendidly," for which Charney was very glad.[37] Working at Chicago on stability, Platzman had discovered that misconceptions could arise if one considered the *shape* of the mean-flow velocity profile as the only controlling factor for stability. After modifying his approach, he had to set aside the calculations on which he and Phillips had worked so hard.[38]

By late 1951, the group's size remained desperately small and almost unable to function. Only two meteorologists (Norman Phillips and Joseph Smagorinsky), a coder, and Charney were working on the project—too few to tackle awaiting tasks. They spent most of their time programming models for the new computer, which was now "physically complete." They broke this large task into several parts as they concentrated on choosing the most advantageous variables for the vertical coordinate in three-dimensional space, the appropriate grid spacing, and the optimum upper boundary layer.[39]

Across the Atlantic, life was very busy at the nascent Institute of Meteorology as its staff members—a team composed of Europeans from several countries as well as American civilian and military meteorologists—prepared their own attack on the problem of numerical weather prediction in early January 1952. A detailed mid-January letter from Bolin filled Charney in on the latest news from Stockholm. He assured Charney that Rossby still "loved him" even though he had not yet answered his last two letters. The Institute had an uncertain financial base, and Rossby was focused on administrative concerns. The Stockholm team members, however, had been busy making tendency computations for Europe and the eastern Atlantic—the same sort of computations that the Meteorology Project had made a couple of years earlier for the United States—and they would be publishing the results in *Tellus*. Team members were trying out a variety of modeling scenarios. Eliassen and William Hubert—a US Navy aerologist studying with Rossby—were applying the barotropic computations to the blocking cases and attempting to find the reasons behind any errors. Meteorologist Karl H. Hinkelmann was working on the three-dimensional model, but would be returning to Germany in just a few days. Icelander Snorre Arnasson was attempting to forecast surface pressure changes using his own technique. Ernst Kleinschmidt was examining potential vorticity in a scenario over the United States. Phil Clapp of the Weather Bureau was making barotropic computations for five-day mean maps. US meteorologist Chester Newton was performing a detailed analysis of the jet stream. Bolin was considering the possibility of using a very large grid to forecast the large-scale flow pattern. He had determined that one had to ascertain a good initial field and then compute mean values so that the few points available were representative. Bolin was also working on a simple three-dimensional model—one with three vertical points—but was not yet ready to discuss it.[40] In short, Rossby's institute was making good progress on numerical model development.

A week or so later, Bolin wrote again. He was anxious to avail himself of an opportunity to go to Princeton at the end of the spring term (1953) and stay until January 1954. With the Swedish Air Force's computer BESK (Binär Elektronisk Sekvens Kalkylator, meaning Binary Electronic Sequence Calculator) being built in Stockholm, he wanted to get up to speed on programming flaws by working with the Princeton group. He also wanted time to visit Chicago, Woods Hole, and MIT (partly because he wanted his new bride to see something besides Princeton). Since his last visit with Charney, he had been working on two- and several-parameter models and had almost finished a paper on the latter. He closed by saying that Rossby hoped to place Phillips in Bolin's position in Stockholm.[41]

The "air mass mixing" of personnel between the United States and Sweden—and between Princeton and Stockholm, in particular—accelerated after Rossby established his new International Meteorological Institute. Rossby's Stockholm group had previously been occupied with conducting small-scale studies for the personnel-deficient Meteorology Project. But by late 1951 the Stockholm group had begun its own work on numerical weather prediction, focusing on eastern Atlantic and western European data. With Rossby arranging the movement of meteorologists across the Atlantic in both directions, the Princeton group's numerical techniques were being aggressively extended and applied outside the United States.

"Johnniac": Alive and On Line

The project's quarterly progress report released in late March 1952 announced a major turning point for the Meteorology Project: the IAS computer—nicknamed "Johnniac" by some—was finished at last. With this milestone, reached 5 ½ years after the project's stumbling beginning, Charney and his team at last turned their attention to using the machine. Although they had considered using their three-dimensional model to test the new computer, they opted to run the older barotropic model first because they already had experience with it. This time they intended to vary the grid size, the time intervals, the relaxation methods, and the numerical storage specifications. The team also planned to use two-dimensional models incorporating essential barotropic features.

During the first run, the team members had an important practical insight: it made more sense to have the machine check itself and stop when it had made an error than to check the output for accuracy at the end of the run. They also streamlined their input technique and found a way to reduce round-off errors. Using the first few integrations of each run to locate machine errors, they would correct the offending program code and run it again. However, during this initial testing stage they did not attempt to draw conclusions from the resulting forecast because of the high probability of error.[42]

Rossby had been in the United States from January until April 1952, but unable to visit project members in Princeton. In a written report, he told them to watch for the needed analyzed maps that were en route via the US Embassy. Eady had joined them in Stockholm and had given an excellent seminar on the 2½-dimensional model. But since BESK was not yet ready, they could only test it with tendency calculations—the same method the Princeton team had used before the ENIAC trials. Finnish meteorologist Lauri Vuorela was visiting for a couple of months, studying atmospheric deformation fields. Rossby was trying to find a mathematician who was trainable in meteorology to help them as soon as BESK was ready, and had already identified a possible candidate.[43]

With progress being made in Stockholm, Rossby wanted Charney to come for a visit so they could share ideas. The frantic pace in Princeton, however, meant that Charney would have no chance to escape. Besides, until he had new products to show from the IAS computer, Charney did not think it worthwhile to make the trip.[44] Rossby understood the inadequate staffing situation facing the Princeton group. He suggested that Charney add both Chester Newton and his meteorologist wife, Harriet, to the team to provide help on the synoptic side. After a year of dealing with theory in Stockholm, the Newtons could be a considerable addition to the project. Rossby suggested that they keep Danish meteorologist Ernest Hovmöller in Princeton until the Newtons returned to the United States. Hubert was working with Eliassen and could be assigned to Princeton upon his return. A Navy officer, he had sufficient "charm and tact" to blend in with the Princeton group without bringing a "military atmosphere." But as far as theoretical meteorologists went, Rossby had no one to recommend: he found the situation in the United States rather desperate. There was no one at the University of Chicago. Rossby also thought that Starr was educating his MIT students in a way that did not match Charney's requirements in Princeton. Hinkelmann might be a possibility. Rossby offered to bring him to Stockholm for "inspection" and determine if he would be a suitable addition to the Meteorology Project.[45] More than 5 years after the end of World War II, the dearth of graduate-trained meteorologists in the United States remained acute.

Harry Wexler visited the Princeton group in mid May for a briefing and a demonstration of the new computer, which was one-tenth the physical size of ENIAC. When working correctly, it could compute a 24-hour forecast in 3 hours. Von Neumann was postponing a dedication ceremony until he was sure all the "bugs" were out of it. Wexler reported back to Reichelderfer that Charney's primary concern remained the lack of personnel, especially synoptic staff, not funding. The Weather Bureau had been providing analysis assistance via WBAN for quite a while, but the increased need for analysis assistance had stretched the bureau to the limit. Namias's extended forecasting section was lending a hand when it could, but it had few slack periods. Smagorinsky had completed his PhD and wanted to return to the bureau, turning down a position with the Air Force's Geophysics Research Directorate. He hoped to introduce electronic computer techniques to forecasters. Closing his report, Wexler wrote: "I am quite impressed by the progress shown by this group and feel that continuation of this progress will go far to promote the *science* of forecasting."[46]

Having discussed the Meteorology Project's progress, Weather Bureau leaders realized that numerical methods would soon shift from research to operations. Reichelderfer recognized their budget forecasts needed to include funding for numerical weather prediction—earmarks for computing and accompanying ancillary equipment, as well as personnel training. The bureau wanted to be "in the forefront in its readiness to adopt improved techniques" by having its employees trained well in advance of operational implementation.[47] Since the bureau had long been criticized for being behind the times, there is little doubt that, with all the publicity and media hype surrounding computer forecasting, Reichelderfer wanted to be ready to move ahead with numerical prediction as soon as it was operationally feasible.

In Sweden, the Stockholm group—having held a conference in early May—reported increasing interest in numerical techniques. The Deutscher Wetterdienst (German Weather Service) had reported on its proposed attempt to compute a three-dimensional tendency field by using ten vertical grid points. Fjørtoft had introduced a graphical technique whereby one could produce a 24-hour forecast in about 3 hours. Conference attendees had strongly supported the prompt introduction of this method to forecast centers. All had been very curious about the status of the Princeton computations, but understood that team members were busy addressing computer problems. The Stockholm group very much wanted to hear the latest news as soon as Charney or Smagorinsky had time to write.[48]

::

During the first few months that the computer was fully operational, the Meteorology Project's members concentrated on testing the barotropic model at the 500-mb level. To adequately test a variety of three-dimensional models, they chose to use the same highly baroclinic scenario for each run: the historic 24–25 November 1950 storm over the eastern United States (figure 6.1). New York City's predicted weather had been for occasional rain and wind followed by some snow. What they got was hurricane force wind and rain, followed by ice and large amounts of snow. Almost 300 people died in 22 states, and damage was estimated to top $400 million. As a *New York Times* editorial mused a couple of days later, "if such a thing had occurred in Russia the Weather Bureau would probably now be on its way to Siberia."[49] Because forecasters' predictions for this systems had been spectacularly wrong—a "bust"—the Weather Bureau had compiled extensive documentation. Therefore, this scenario was a logical choice to see how well the computer runs would compare with observed data. The November 1950 storm was also interesting because its rapid development involved conversions of potential energy to kinetic energy—something that models would have to be able to handle effectively if numerical weather prediction were to become operational. Models that could only predict run-of-the-mill low-pressure systems moving across the

Figure 6.1
The 1950 Thanksgiving holiday storm that dropped these large snowdrifts in Massillon, Ohio, continued to the East Coast and strengthened, catching forecasters by surprise. (courtesy of Collection of the Massillon Museum)

continent would not be a significant improvement over what experienced forecasters could do without computer assistance.

In spring 1952, project members decided to compare two models, the equivalent barotropic and a simple baroclinic, against a barotropic model. The barotropic model only predicted flow at the 500-mb level because it contained no vertical structure. And since this model did not make potential energy available for conversion to kinetic energy, it could only predict a redistribution of initial kinetic energy. The equivalent barotropic model would serve as the control. When the computer operated at full capacity, the 24-hour forecast required a run time of only 90 minutes—10,000 times faster than a person doing the same calculations with a desk calculator, which would take 8 years of 40-hour weeks and provide a very old hindcast of dubious value. An increase in program efficiency could reduce the 24-hour forecast run time to 10 minutes. In all, the team made twelve runs: six 12-hour and six 24-hour barotropic forecasts. As expected, the barotropic model did not fully predict the storm's extremely rapid and intense development. The group decided to overcome this by creating a simple baroclinic model that would take the three-dimensional character of the atmosphere into account. Consisting of barotropic layers at 700 mb (10,000 feet, 3,000 meters) and 300 mb (30,000 feet, 9,000 meters), they reduced the time step to 30 minutes to avoid computational instability. At full speed, the computation time was 2½ hours for a 24-hour forecast. Since the machine typically ran at half speed, it took 5 hours. As was hoped and expected, the baroclinic model performed better than the barotropic model with the 24-hour forecast, showing less degradation from the 12-hour forecast. The group was thus encouraged that the baroclinic model might give reasonably good forecasts out to 36 or even 48 hours. Because the two-layer model included a vertical component, it could also predict cloudiness and precipitation, which depend on rising air. The team members concluded that the baroclinic model output agreed fairly well with the observed atmospheric state.

The two-layer model was only a small step on the path to operational numerical weather prediction, not the destination. Because the two-layer model could not account for horizontal variations of static stability, the team members would need to create a model with at least three layers. This would entail overcoming many theoretical and programming difficulties, and making simplifying assumptions that would allow the team to develop computer-solvable equations. Once the team members had completed the programming, they had to await the arrival of a magnetic drum that would provide the additional memory needed for coding and running the model. Eventually, they hoped to include frictional effects at the earth's surface, non-adiabatic (heat-exchange) effects, and large-scale orography.[50] Although it had only taken a few months to advance their model from the barotropic version to a simplified baroclinic version, the dream of including major frictional and thermodynamic effects in atmospheric models would not become a reality until the late twentieth century—when computer capacity was sufficient for such a task. With each new generation of computers, the models would advance, albeit slowly.

How did the model output stack up against the prognoses produced by experienced forecasters? Although project members had not made detailed comparisons, the model output and hand-produced prognoses appeared to show comparable accuracy. They thought subjective forecasts had not improved in the previous 20 (or even 50) years—a widely held opinion among meteorologists. Although model output was not yet superior to subjective forecasts, at least the models could be improved incrementally as theory, mathematical techniques, and computing power advanced. Due to the project's work, efforts at numerical weather prediction were expanding around the world as the Swedes, the Norwegians, the Danes, the Japanese, the British, and the Germans conducted related research. By 1952, prototypical techniques for numerical weather prediction had gained the respect of meteorologists of all backgrounds. They accepted that scientific gains could only come from the application of computerized numerical techniques. Without them, potential advances made possible by increased surface and upper-air observations would come to nothing because the overwhelming amounts of new data could only be digested by electronic computers.

Additional efforts in anticipation of operational numerical weather prediction included pursuing aspects of general atmospheric circulation related to longer-range forecasting. Team members studied the influence of large-scale longitudinal asymmetries in heating on the mean seasonal flow pattern, showing that heating differences demonstrated the most important effect on normal lower tropospheric patterns. They also examined the role of unstable baroclinic disturbances in maintaining the energy balance. That study indicated that large-scale baroclinic disturbances converted sufficient potential energy to kinetic energy to balance the loss of energy due to friction, and at the same time transported enough heat poleward to balance the net radiation loss.[51]

Despite a chronic personnel shortage, the Meteorology Project was still able to make significant gains between the first ENIAC expedition and the period just following their tests on the new Princeton machine. Project members tried models, comparing their output against analyzed data, and modifying them for the next run. They examined barotropic and baroclinic models along with models in two, two and a half, and three dimensions. But as Rossby had long maintained, to sell the meteorological community on the effectiveness of numerical weather prediction, and simultaneously advance model development, the project members would need to put their models on the line—the one line that mattered, the operational line.

The Air Force Re-Engages: Thompson on the Move

While Jule Charney and his group had been sorting out ENIAC results at the end of 1950, Philip Thompson had been preparing for an early 1951 trip to Europe. The purpose, in his own words, was "to establish scientific contact with several meteorological institutes in Western Europe, observe research in progress at those institutions, estimate their capacities to expand the scope and scale of their present research programs, and if it appears desirable, to initiate such preliminary negotiations as are necessary to arrange for partial subsidization of research in meteorology and weather forecasting."[52]

Why was Thompson making these contacts? Because he had convinced the Air Weather Service that there was not enough research capacity—manpower or facilities—in university meteorology departments in the United States to fulfill the Air Force's needs. The demands of Navy and Air Force research contracts had so stretched the capacity of American meteorology departments, in his view, that a national emergency would find them with insufficient faculty and staff to meet mobilization-driven training. Therefore, it was important to assess opportunities to expand their research contracts to western European countries.[53]

Writing to his contact stationed with the US Air Forces in Europe (USAFE), Ramstein, Germany, Thompson further explained that he needed to find out what European meteorological institutes were doing so that he could work toward "meshing" the Air Force's meteorology research program in the United States with those of his western European counterparts. He wanted to "see where a few hard US dollars would do the most good." Therefore, Thompson planned to visit the British Meteorological Office and the "Brunt school" (Imperial College) in the United Kingdom, the Oslo Institute, the Stockholm Institute (Rossby), and five German institutes.[54] He informed Arnt Eliassen that the "[main] purpose of this trip and of my visit to you is simply to get a general idea of the sort of problems which preoccupy the European meteorologists and to absorb something of their viewpoint."[55]

At each stop on his European tour, Thompson analyzed staffing, workload, funding, and personnel available for contract research. Thompson also pondered the legality of using a host nation's funds for facilities and US funds for research. He wanted to determine if this arrangement would be legally and practically possible before requesting formal contract proposals.[56]

By spring 1951, the Air Force had established the Office of Air Research to compete with the Office of Naval Research for the hearts and minds of scientific researchers. However, the OAR was not in a position to finance basic research in the manner of the ONR, which was also moving toward funding more applied research. Therefore, Thompson advised his friend and colleague, Chankey Touart, a doctoral student at NYU's Institute for Mathematics and Mechanics, that he should not expect the ONR to continue supporting the Institute's mathematical research at its previous level. Thompson argued that the ONR had so taken advantage of its newfound freedom in 1946 to pursue basic research that it had "overshot" and reached an "untenable position." No longer able to justify its research strategy to "Battleship Admirals," the ONR was moving rapidly away from basic into applied research. In Thompson's opinion, this move would affect support for meteorological and geophysical projects. Organizations desiring to pursue basic research projects could improve their funding potential by disguising them as applied projects.[57]

Thompson left his post as Chief of the Atmospheric Analysis Lab at the Geophysics Research Directorate on 1 August 1951, spending the rest of the year and into spring 1952 finishing his PhD at MIT. As Charney and his group were preparing to run models on the IAS computer, Thompson received a request from the Air Weather Service to submit his "dreamsheet" (preferences) for his next assignment. Thompson did not want to find himself relegated to an operational post (in his case, a weather forecasting office) after spending his entire career in research and development. On the contrary, he intended to take whatever steps were necessary to ensure his continued leadership in, even control of, the Air Weather Service's program for numerical weather prediction. Therefore, he made clear in his response that to his knowledge, the Research and Development Command had already requested, or would shortly request, his assignment back to the Geophysics Research Directorate. As he pointed out, he had spent virtually his entire career in the Air Force working in research and development, and that certainly seemed to be the best use of his interests and abilities.[58] However, Thompson must have been feeling less than fully confident of this assignment. Jumping his chain of command, he sent a query letter to Sverre Petterssen, Director of Scientific Services for Headquarters, Air Weather Service. Thompson and Petterssen had recently attended a conference on general atmospheric circulation. He told Petterssen that he had been thinking about how numerical prediction could be exploited and wanted to make sure Petterssen understood his thoughts. (Thompson, showing that he wanted very badly to be reassigned to the Geophysics Research Directorate's meteorological research arm, despite being an Air Force major, wrote directly to Petterssen—a clear breach of the command structure. Clearly, he had a direct supervisor—no one in any military service is a "free agent"—and it was not Petterssen.)

In his letter, Thompson sought to persuade Petterssen that current numerical methods already were producing forecasts equivalent to existing, traditional hand-drawn forecast maps. This was patently untrue. The prognoses popping out of the IAS computer were certainly meteorological (they had the curved lines and closed circulations one would expect to see on a weather map), but the models still did not handle rapidly changing systems well. The intensity and position of high-pressure and low-pressure areas were not accurate enough to use for forecasting without additional human-created meteorological products. However, had Thompson not made this claim, his argument for operational numerical weather prediction would have fallen apart. Meteorology Project members and their supporters would later make a similar argument to obtain necessary funding for operational numerical weather prediction. Nevertheless, numerical products would not be equivalent to forecaster-drawn maps until the 1960s, and even then they met only a minimum threshold of acceptability.

Continuing with the premise that the computer-generated weather maps were the equivalent of forecaster-produced maps, Thompson argued that because these numerical forecasts were completely objective, one would at least know where to look to determine why a bad forecast had failed. He anticipated that numerical forecasts would continue to improve as modelers found and removed errors. His own doctoral research, Thompson pointed out, was an effort to "weld together the mechanical and thermodynamical aspects of the prediction problem" and that in the future numerical models would predict temperature and vertical motion fields in addition to pressure and horizontal wind fields. When that time came, true objective weather forecasting would be at hand. Further, Thompson thought that waiting for something better was "psychologically untenable" for those working in numerical weather prediction. He argued that to sustain an operational organization for numerical weather prediction one would have to develop the models, build the equipment, find suitable physical facilities, and secure well-trained specialists. Those specialists did not exist, at least not in sufficient numbers, and would need to be trained. By Thompson's estimate, if the Air Weather Service started immediately, 2 years would pass before they would be ready with an operational organization. Given the incremental progress in numerical methods, Thompson thought it was safe to pursue the creation of an operational center for numerical weather prediction immediately. It was a bold gambit, but entirely consistent with Thompson's earlier career.

Thompson did acknowledge potential obstacles, especially a complete lack of trained manpower. Where could sufficient numbers of people receive training in numerical weather prediction? Thompson maintained that the Princeton group should not undertake a training mission because it would detract from its research mission. Even at the Geophysical Research Directorate in Cambridge, only a few people were familiar with numerical techniques and the theory behind them. Unfortunately for the United States, he noted, virtually everyone who was active in the field had come from Scandinavia: Eliassen, Fjørtoft, Einar Høiland, Bolin, and "of course Rossby." (So, of course, had Petterssen.) Thompson's suggestion was to "move Mahomet to the mountain"—that is, send people to Stockholm for training. Conjecturing that it would take Rossby at least a year to package a course of instruction, Thompson thought this alternative would work just fine, because it would coincide with the completion of Stockholm's new computer.

Exactly what did Thompson want? He clearly wanted to go to Stockholm, even if it was not good for his career. (In most cases, accepting too many assignments away from the operational forces reduces a military officer's promotion opportunities. Thompson had never really served in an operational job.)

Timing this letter to reach Petterssen the day before he left for Scandinavia, Thompson asked Petterssen to intervene with Rossby on his behalf.[59] The same day, Thompson sent a letter to Charney, enclosing a copy of his letter to Petterssen. He asked Charney to "show the letter to no one" and to discuss the contents only with Rossby. However, Charney was not supposed to let Rossby know that the contents came from his letter to Petterssen. "I am sorry to appear so conspiratorial about this," Thompson continued, "but my position would be extremely awkward if all the gory details were known to the people at Wright Field [the home of Thompson's superiors]."[60] "Extremely awkward" was a major understatement. Thompson's very career could have been in jeopardy if his superiors had discovered that he was working behind their backs and outside the chain of command. He had to hope that Charney would not inadvertently expose his subterfuge.

Thompson's letter to Petterssen had indicated that the time was ripe for pursuing operational numerical weather prediction. But in his letter to Charney, a disingenuously irate Thompson attempted to put Charney on the defensive over a perceived deception: that project members had not kept him fully advised of how close they were to implementing some sort of operational organization for numerical weather prediction. Thompson was accusing Charney of conspiring with others—presumably Rossby, von Neumann, the Navy, and the Weather Bureau—to keep Thompson from taking part in operational numerical weather prediction. Furthermore, numerous options no longer existed since so much progress had been made toward operational numerical weather prediction. In fact, just one appeared viable: a joint organization manned and funded by all the weather services. Thompson had opposed a joint venture, but upon realizing that a jointly manned and funded activity might form whether he wanted it or not, desperately claimed that he did not want to be left out. Of course, that was not what Thompson had told Petterssen, and Charney knew it. Thompson was trying to play on both sides of the fence and claim a spot with the victor. Hedging his bets, he told Charney he wanted to go to Stockholm. Would Charney, he asked, please pass that piece of intelligence on to Rossby?

A main point of Thompson's earlier campaign against a joint operational unit had been to dispute an important assumption of the Princeton group: that to make an accurate prediction, numerical modelers had to address three-dimensional motions in the atmosphere. On the contrary, Thompson thought, it was possible to deal with baroclinicity in general by the "mathematical artifice" of vertical integration. Intending to visit New York after Easter (13 April 1952), he hoped to visit Charney to discuss his ideas further.[61] Such a comment suggests that Thompson was not seeing Charney every "two to three weeks," as he later claimed in recounting his recollections of the Princeton Meteorology Project.[62] Indeed, it appears that Thompson was not in regular communication with the Princeton group. The next day, a still boldly calculating Thompson sent a letter directly to Rossby with this opening paragraph:

We were in some hopes that you might turn up at the general circulation conference at MIT last week, but learned that you were returning to Sweden on Wednesday, nursing an incipient cold and otherwise worn by winter travel. (How you are able to withstand the pace of commuting between Chicago, Stockholm, and Princeton is a mystery deeper than the confluence theory.) At any rate, since it is unlikely that I shall see you before your next visit, I felt that I should write to state my position as clearly as is possible under the present circumstances.[63]

Abandoning military protocol, Thompson pushed on: he told Rossby how much he wanted to join his Stockholm group. Thompson understood that Rossby might be concerned about a military person's injecting "a completely alien outlook," but wanted him to know that his only interest was in research, not in a series of administrative staff jobs. (Since William Hubert was already in Stockholm working with Rossby, and Daniel Rex had completed his PhD under Rossby in Stockholm, Thompson likely realized that his military affiliation was of no concern to Rossby, making his comment disingenuous.) Thompson declared that he was "looking forward to a visit to Sweden as a sort of preview of the meteorologists' Valhalla (which I am sure must be somewhere in Scandinavia) where the ghosts of old heroes still mingle with younger warriors." Thompson was less concerned with ghosts than he was perturbed that the forward movement in operational numerical weather prediction was a fait accompli and that nothing he did would affect the outcome. He was determined to go to Stockholm, no matter how that would be viewed by the Air Weather Service leadership.[64]

Nothing came immediately from Thompson's exertions. He soon received a response from Petterssen, who was glad to know that Thompson agreed with him on "all essential points"—a fascinating statement, since Petterssen was the one in charge. Petterssen was expecting to see Rossby soon and told Thompson he would get back to him later in the spring.[65] Rossby, meanwhile, had received Thompson's "flowery letter" and was not quite sure what he wanted. Surmising that Thompson wanted to join his group, Rossby told Charney that it was fine with him, but that he would talk to Petterssen, soon to arrive in Sweden.[66]

Thompson's letters strongly suggested that he was very much interested in being a team player with the rest of the people working on numerical weather prediction, that he *really* wanted to go to Stockholm to study with Rossby, and that he was willing to pull any strings possible to get what he wanted. Yet it seems evident that throughout most of his tenure at the Cambridge center Thompson had been actively seeking to steer Air Force meteorology's research agenda. The trip to Europe by so junior a person to evaluate possible research contract opportunities, the astute assessment of research differences between the Air Force's and the Navy's research arms, the drive to secure an operational run of his models, and the behind-the-scenes, "don't tell anyone about this letter" maneuverings for his next assignment all point to a scenario that was not exactly what it appeared to be to any single individual involved with Thompson. But over the next few months, a very different scenario would be played out—not the "team player" scenario, but one that would quickly force the hand of everyone who had been involved in the Meteorology Project. Indeed, all of Thompson's correspondence appears to have been a rather elaborate cover for what Thompson really wanted: a shot at controlling operational numerical weather prediction.

Who Would Control Numerical Weather Prediction?

While Thompson was making overtures to Rossby and Charney, Chankey Touart—Thompson's replacement as Chief of the Atmospheric Analysis Laboratory of the Geophysical Research Directorate in Cambridge—was writing to the Commanding General of the Air Research and Development Command to propose the establishment of a Numerical Prediction Project (NPP) with Philip Thompson as its leader. Almost certainly, Thompson was behind this initiative.

Touart explained that two groups were working on weather forecasting using electronic computers: the IAS group (focusing on basic research and a thorough, systematic exploration of fundamental problems) and the Geophysics Research Directorate group (led by Thompson until the fall of 1951, and focusing on applied research and concentrating on the development of mathematical models and their exploitation with a more immediate concern for practical forecasting).

Work on numerical weather prediction had focused on the creation of "idealized models" of the atmosphere that would be solvable by an electronic computer. The Geophysics Research Directorate group had focused on a linearized, two-dimensional, adiabatic, quasi-barotropic model that assumed an average wind in the vertical, no energy input to the atmosphere, and that potential energy is only minimally converted into kinetic energy. Nonetheless, test runs had shown the model could produce a 24-hour forecast of 500-mb-level winds commensurate with those produced by an experienced forecaster. The model broke down in cases that were more likely to cause a subjective forecast to bust: cases in which there was a new system developing.

Touart then described the IAS model as a "non-linear analogue" of the Geophysics Research Directorate's own model. As far as he knew, once the members of the IAS team had evaluated their new computer, they planned to make forecasts based on a three-dimensional version of their model. Touart did not have much on which to base a comparison of the models, but he guessed "on physical grounds" that the IAS model would not produce results radically different from the GRD's except that the IAS model would produce an actual, not an averaged, wind.

The Geophysics Research Directorate thus planned to introduce topographic effects, to prepare experimental 48-hour forecasts, and to use their current computing machines (then just coming into use) to create 24-hour forecasts on a "production" basis (presumably meaning in real time). Thompson, Touart went on, was crucial to the success of this work: his current research was focused on the "strong-development" problem, which appeared to have practical applications. Additionally, Touart suspected that Charney was ready to take the IAS project from experimental forecasting to operational implementation. Therefore, it would be of the greatest advantage to merge the operational and research arms of numerical prediction into one Numerical Prediction Project headed by Thompson, whose "background, capabilities and burning aggressiveness" were required for the "*immediate* success" of this project. Touart requested Thompson's assignment back to the Geophysics Research Directorate to head up the proposed Numerical Prediction Project. In the meantime, Thompson would be assisting "unofficially" in getting that project established.[67]

Within a few days of Touart's letter, the internal newsletter *GRD Spectrum* published a small article on the "pioneers" in the Atmospheric Analysis Laboratory, who would, "for the first time in meteorological history, as far as we know," produce an upper-level wind forecast by machine methods during the forecast period. Lou Berkofsky, a civilian meteorologist working for the Air Force, would direct the running of Thompson's model on real-time data. They hoped to have the forecast out 12 hours after the first PIBAL (pilot balloon, used to obtain wind velocity aloft) report came across the wire.[68] If Thompson had been surprised that Charney's group was nearly ready to pursue an operational venue, he certainly cannot have been in the dark about what the Atmospheric Analysis Laboratory was doing with his model.

Thompson maintained a low profile about his involvement with the proposed Numerical Prediction Project until early July 1952, when he began sending out query letters to potential staff members. Since all the letters were essentially the same except for the opening personal paragraph, the one sent to Chester Newton in Stockholm serves as a good example. Thompson began by providing background information: several institutions in the United States and overseas, including the Geophysical Research Directorate, had groups working on numerical weather prediction. Unlike the other groups, the GRD had been "pursuing a long-range program" that would ultimately move into semi-routine numerical predictions. It was now time to exploit and apply the methods that had been created. Thompson added that he believed Rossby would "attest to this." (This comment must have been added for Newton's benefit, since it did not appear in the letter to a UCLA staff member: Thompson, wishing to connect himself to Rossby, was counting on Newton to discuss the contents of the letter with Rossby.)

To speed up the process, the GRD planned to bring the research and operational meteorologists under one roof to facilitate a two-pronged attack on the problem. In case Newton might be wondering why the Air Force would choose to do this on their own, instead of jointly like the Meteorology Project, Thompson reassuringly declared that it would take 12–18 months just to get through all the administrative negotiations and hassles. Therefore, it seemed best for the Air Force to set up its own outfit and "invite" people from other weather services to participate. The target date was 1 February 1953. After asking Newton about his interest in accepting an appointment, Thompson inserted a disclaimer that he had no official status in relation to the project although it was "tacitly assumed" he would be its director. "Meanwhile," he wrote, "I am acting as unofficial head of an almost non-existent project—as an entrepreneur, if you like."[69]

Writing to Charney a week later, Thompson tweaked him for not responding to the accusation, leveled in his 31 March letter, that Charney was involved in some sort of conspiracy against him. Thompson then explained that his situation had changed, and discussed the GRD's plans to establish the new research/operational project in Cambridge. He argued that the development and application of numerical methods would proceed most rapidly if all aspects of the problem could be attacked "simultaneously and under the same roof." The "roof" that Thompson proposed covered the Air Force's Geophysics Research Division in cooperation with the Air Weather Service. He listed four subsidiary aims: to develop, test, and evaluate new methods of numerical prediction; to devise high-speed computational and data-processing methods and equipment; to prepare and send numerical products on a "semi-operational" basis; and to provide experience to personnel who would then move to field units.

Apparently, Charney and Thompson had already discussed how a future operational unit might look, because Thompson made it clear that he knew Charney did not support having such work done by a single entity. However, if numerical weather prediction were handled only by the Air Force, there would be less bickering and they could get started that much faster. The new research group would include six people: Thompson, a civilian who would replace Thompson when he transferred, two meteorologists, and two junior meteorologists. But Thompson needed to find some researchers to join his group. Did Charney know of anyone who would be available? In particular, he intended to sound out Phillips—already under contract to Charney's group in Princeton—and hoped that Charney would not regard it as "proselytizing." And one last thing: Thompson wanted to know if Charney had any ideas about "inexpensive and efficient computation devices."[70]

Discerning Thompson's vast ambition, Charney was livid. He telephoned Wexler—a relatively rare occurrence—and gave him the gist of Thompson's letter. Charney insisted that he did not like the idea of cutting the Weather Bureau out of numerical weather prediction, in view of all the help it had provided over the lifetime of the Meteorology Project. Furthermore, Charney considered the bureau to be *the* federal meteorological service—not the Air Weather Service. He was sure that von Neumann was going to be "greatly disturbed" when he got the news. Petterssen obviously was involved, since the GRD and the Air Weather Service were establishing the new unit together. There were also rumors about forming a unit in Washington. That would mean four proposed units: one at Chicago, one at Cambridge, one at Princeton, and one at Washington.[71]

The very next day, Wexler was on a train to Cambridge to meet Chankey Touart and Helmut Landsberg (a climatologist who headed the Geophysical Research Directorate). Backpedaling, they denied that Thompson's letter had any official status—after all, he was simply an MIT student and not a GRD member. Touart did admit, however, that the letter was written with his knowledge. The descriptive points about the unit? They were from a memo Petterssen had sent to General Oscar O. Senter of the Air Weather Service. But the proposed unit was not nearly as pretentious as Thompson had made it out to be. It was just an extension of the small extant numerical weather prediction unit within the Atmospheric Analysis Laboratory. They planned to increase the number of employees to 24, and to use punch-card machines to find "analytic solutions for a linearized model in 24-hour jumps." At that point in the conversation, Wexler, Touart, and Landsberg called Jule Charney. When asked about the linear models, Charney maintained they were "hopelessly out of date." Therefore, Charney believed it would be a complete waste of time, money, and manpower to test them. Disturbed by this conversation and by its larger implications, Charney decided to call a meeting in Princeton. Thompson's letter had brought the matter to a decision point, and it was necessary for representatives from all the weather agencies to gather and develop a plan. Landsberg thanked Wexler for bringing the matter to his attention.[72] If Landsberg's comment was genuine, that Thompson and Touart had together jumped way past the chain of command on an official matter was made clearer still. The situation had been bad enough when Thompson wrote directly to Petterssen to request his help in getting to Stockholm, but when Touart skipped past his immediate superior—Landsberg—and went directly to Air Weather Service with his proposal, the potential for professional disaster and embarrassment had been greatly magnified.

Wexler shared his more personal thoughts on selecting an operational model with Francis Reichelderfer. In military terms, they were walking through a minefield and would need to be very careful. The decision was being influenced by "strong ambitious personalities and presumably interservice rivalries." It would be difficult to select any one model to be operational since each had its strengths and weaknesses. While the researchers may have been looking for some kind of perfection, the forecasters were looking for a product that would help them out in the near term. They would need to be careful. Any quick moves into operational numerical weather prediction could discredit the entire concept even if "obsolete" models were the ones used.[73]

Charney recognized that there were not enough appropriately trained people to staff several centers for numerical weather prediction : the long-running dearth of professional meteorologists remained troubling. If they did not work together, then they risked failing. Having worked together from the beginning, it was unconscionable for one group to engage in empire building while shutting out the people who had brought numerical techniques to the verge of operational status. In an unsuccessful attempt to bring these matters to Thompson's attention, Charney's response was extremely blunt: "I saw no reason for answering your earlier letter with its fantastic imputation of ulterior motivations. Conspiracies are foreign to my nature, besides they take too much time." Continuing, Charney argued that it was unwise to shut out the other weather services. The Air Force, the Weather Bureau, and the University of Chicago were already establishing a program in Chicago that would be performing the same kind of work that Thompson was proposing for the Geophysics Research Directorate. Charney did not think that the GRD was better equipped than any other to become an operational center.

With so few qualified meteorologists available, "if you succeed in obtaining good meteorologists who are already trained in numerical forecasting," Charney wrote, "will you not be robbing Peter to pay Paul? What is needed now more than anything, I repeat, is to train people."[74]

Furthermore, Charney declared he was "astonished" that Thompson had not consulted with him or von Neumann about these plans. While they were not "building empires" at the IAS, they did have a one to two year lead on everyone else, and were glad to share their experience. Particularly irksome was that the GRD now provided half of the Meteorology Project's funds: the IAS was openly competing with its own contractor. Charney made clear that Phillips was staying with the Meteorology Project—they had contracted with their funding agencies for 4 years just so they would have the long-term stability they needed to hire people and keep them.[75]

But Thompson seemed little dissuaded by Charney's letter. On 1 August 1952, he sent a letter to Bert Bolin in Stockholm asking him to join the GRD group. Bolin declined.[76] So did Newton, who thought that any operational unit needed to be "joint." Newton's other concern was more fundamental: too rapid a move into operational numerical weather prediction could leave them with forecasts of lesser quality than were available by subjective, hand-drawn techniques. If that happened, many people could be inclined to kill operational numerical weather prediction aborning.[77] Morton Wurtele of UCLA was more interested in Thompson's offer. Besides wanting to know his possible civil service employment grade, Wurtele was concerned about how he could get his work published if he accepted a position.[78] Thompson clarified for him that the purpose of the group would be to develop the theory on which numerical weather-prediction methods were based and to apply a theory that Thompson himself had just worked out.[79] This inadvertently revealing comment suggests that Thompson had no intention of using any other group's models in his new unit.

Later in August, Charney sent a copy of Thompson's letter (minus the first "conspiracy" paragraph) to von Neumann along with his comments. "This is empire building in its crudest form," Charney wrote, clearly appalled that Thompson's proposal had been made without consulting either the only experienced group working on numerical weather prediction or the Weather Bureau itself. At the same time, he realized that they were not going to be able to stop Thompson's move into pre-operational research and training for numerical weather prediction.

The only thing to do, Charney believed, was counterattack. Numerical weather prediction was universal, just like other weather-related subjects and undertakings, he argued. No single weather service should have a monopoly. Just as the joint WBAN analysis and forecast center handled these tasks for all three weather services, so a similar venue should be created for numerical weather prediction. They would need trained personnel to launch such a group, but there was nowhere to train anyone except "in-house" because they were learning as they went along. Even with trained personnel, they would still need research and development on communications, data handling, and objective data analysis that could not really fit under a basic research umbrella. In fact, Charney noted that such research did not belong to any one group or person. Although his Princeton group could train a few people, they really were not equipped to take on that task. An entirely new group—preferably located nearby—was needed to work with the original team members. Or they could set it up with the University of Chicago group, a joint venture under Petterssen's direction. However, Petterssen appeared to be backing Thompson's new unit—a group whose members had to be Air Force personnel. Although the Chicago group and the Weather Bureau were committed to continuing their close working relationships with the Meteorology Project, Thompson had given every indication that he would not maintain close ties to the Princeton team. Charney reported that Wexler was "equally indignant." Charney clearly did not trust the GRD or Thompson. He strongly urged von Neumann to join him in taking the initiative to put operational numerical weather prediction in the "proper hands."[80]

In his correspondence with Charney, Thompson just as adamantly maintained that the Air Force had the right to go it alone for military reasons. Although now claiming the project was a "stop-gap" measure, in his opinion the necessity to support short-range military objectives meant it was "neither necessary nor desirable" to set up a joint operational group. However, Thompson was willing to let the other weather services take part as long as the Numerical Prediction Project was not "too seriously affected by the attendant divisive forces." Likening Charney's version of the call with Wexler, Touart, and Landsberg to the "parlor game called Telephone," Thompson maintained that Touart had told Charney that they intended to test a number of new models and not just Thompson's. If so, then both Wexler and Charney misunderstood Touart, for both were convinced that Thompson's new group would only test Thompson's model.

In his subsequent response, Thompson backed away from his comments about training, implying that Charney himself had tried to put distance between the Princeton group and any operational venture. Knowing that was not the case, he promised to "belabor and besiege" Charney for assistance. He appreciated the offer of training assistance even while commenting on

Charney's "explosion," which had come as "a surprise to everyone" when the topic had been discussed the previous week.[81] Clearly this letter was not meant to invite cooperation.

Von Neumann and Charney, understanding the situation better and better, did not wait to consolidate their position. They called a meeting for 5 August 1952 in Princeton. Reichelderfer sent letters to the leaders of the Navy and Air Force weather services, asking them to attend because there was not much time before the congressional budget hearings. If they were going to plan for operational numerical weather prediction, they needed to develop their justification. The letter included a discussion outline prepared by the Princeton group—probably Charney under the circumstances, although that is not explicit in the letter—which set out four stages that needed to be addressed en route to operational reality: preliminary training, pre-operational research and development (objective analysis, communications analysis and development, forecast evaluation, deductions of weather from numerical forecasts of meteorological variables, derivation of small-scale phenomena from large-scale predictions), and direct preparation for and operation of a numerical forecast service.[82] An earlier historical account maintains that the meeting was called quickly because of concern over the budget hearings alone,[83] but that is highly doubtful. Until Charney, Wexler, and von Neumann were confronted with Thompson's brazen attempt to take the "joint" out of numerical weather prediction, there had been no discussion of having a meeting to provide ammunition for a budget hearing. In any case, the meeting would not have included everyone who had ever been concerned with numerical weather prediction if it were just a Weather Bureau concern. Clearly, Reichelderfer was undertaking a little ruse of his own to get everyone to Princeton and to force a decision to keep operational numerical weather prediction a multi-agency program.

Wexler represented the Weather Bureau. The Air Force sent representatives from the Air Weather Service, the Air Research and Development Command, and the Geophysics Research Division. Thompson was conspicuously absent, but Touart and Petterssen attended. The Navy sent representatives from the Office of Naval Research and from Naval Aerology, Daniel Rex representing the latter group. Princeton staff members rounded out the attendees.[84]

After reviewing the progress made in numerical weather prediction, von Neumann made "inferences" based on what the Princeton group had learned. He assumed a general baroclinic model providing a forecast of at least 36 hours would be the first practical forecast. He noted that there would be three basic steps to any such forecast (input, actual computing, and output), the combination of which should take 12 hours to complete. When programmed for speed considerations, he anticipated the computations would take 4 hours. The other 8 hours would be tied up inputting the data.

Furthermore, von Neumann identified two challenges to overcome: education and technology. The first was a challenge because there were very few people who had both synoptic meteorology and mathematics backgrounds to supervise and operate the program. With an intense program, he thought they could get people trained in about 3 years. Technology challenges included having a machine that was in "perfect condition" and in working order at any time. Petterssen noted that the machine would be "idle" most of the day, but that the "idle" time would be required for maintenance. Von Neumann envisioned that one-third of the day would be devoted to preventive maintenance, one-third to test runs, and the rest for both operational and research runs.

Petterssen questioned the times for input, computation, and output. However, Charney thought they were being sufficiently conservative for an initial attempt. Eventually they would be able to save time in the input and output sections.

Another discussion concerned geographic coverage for numerical forecasts. Von Neumann thought that available machines could cover the United States, but even that geographical expanse might be too large for the 36-hour forecast given memory limitations. If they extended the area to range from Japan to eastern Europe—a fourfold increase—then they would need a much larger machine. One might be ready in 5 years. Von Neumann argued that it would be best to show that numerical weather prediction was viable before requesting additional data that the models would require to fill gaps in coverage.

The Weather Bureau representatives (Harry Wexler and Joseph Smagorinsky) commented that it took 7 hours from data receipt to facsimile transmission of the prognostic chart with current subjective, hand-drawn methods. A question that remained unstated, but had probably occurred to everyone there, was "How does it improve the situation to take an additional 5 hours to get the product out?" Wexler reported that bureau forecasters valued numerical weather prediction for time periods beyond 36 hours because they could already produce accurate products for the 24–36-hour range. However, the general group consensus was that it was "too early" to lead people to think that longer range forecast reliability would be improved. Charney also argued that the barotropic forecasts, as then run, were not as good as a subjective forecast, but preliminary work on baroclinic models showed a promise of improvement.

The other major concern—training of personnel—also had to be addressed, and quickly. In its chronically undermanned state, the Meteorology Project was probably not the best training venue. However, its members did have more experience than anyone else. Therefore, the project members were willing to take in three trainees at a time. The attendees recommended that each of the three weather services send one person to Princeton to work with the project members. Those selected would need to possess most of the required qualifications: solid synoptic meteorology background, mathematics and physics expertise, and a limited need for a refresher course in meteorological theory. Others, including synoptic forecasters with some theoretical background or those with little or no meteorological knowledge, but solid in mathematics and/or physics, were thought to be best trained in a university setting. The University of Chicago had a program suitable for synopticians needing theoretical work. According to von Neumann, Chicago needed to attract more people from applied mathematics and physics into the program to ensure its long-term success. Petterssen noted that weather forecasting did not attract "many theoretical minded people," but believed that as numerical forecasting became a stronger element, more of those kinds of people would be willing to join the discipline.[85] At that moment, the Geophysical Research Directorate's Touart introduced Thompson's personal statement about his plan for a parallel group moving forward with a two-dimensional model. The statement reflected the message sent to Charney, which had been the impetus for their current meeting. Thompson claimed that it was "wasteful" to pursue a three-dimensional model of the atmosphere when his two-dimensional model worked just fine. With a modest investment, they would be able to turn his two-dimensional model into an operational model suitable for routine uses. Thompson maintained that "immediate military needs," which were distinct from the needs of the populace at large or of the research interests of the scientific community, dictated that they work on a model (his) that would be operational in 2 years or less. Thompson's plan would be a short-range program to produce the best model for military purposes within 2 years, while a long-range program pursuing the best possible model could still be undertaken for general use.[86] But short of two dismissive comments by Charney and von Neumann, there was absolutely no discussion of Thompson's proposal. Apparently even the Air Force representatives were distancing themselves from Thompson, even though they were not backing a joint venture.

In the end, the attendees agreed on only a few points. Each service's representative volunteered to provide a trainee to Charney's group. (Owing to other commitments, it appeared that 1 December 1952 would be the absolute earliest arrival date for anyone.) The Navy's representative expressed support for a joint venture modeled on the WBAN. The Air Force's representative declined to comment on the joint venture and indicated his service would continue its own project at the Geophysical Research Directorate while continuing support for the Meteorology Project.[87]

What is most apparent was that any attempt by the Air Force to claim numerical weather prediction for itself was doomed. In his attempt to control numerical weather prediction, Thompson had overreached, and, temporarily at least, those involved in the Air Force's numerical weather prediction were no longer trusted by others in the field. Despite the claim of Air Force representatives that they would proceed alone, what had started out as a joint project would remain a joint project.

Nevertheless, the conflicts inherent in going "joint" were just beginning.

7 ∷ A Changing Atmosphere: From Developmental to Operational Numerical Weather Prediction (1952–1955)

"Electronic 'Brain' Planned to Forecast the Weather," proclaimed a Science Service article splashed across page 12 of the 12 August 1952 *Boston Daily Globe*. Appearing a scant week after the Princeton meeting that had addressed future operational possibilities, the story explained how computers would be making Weather Bureau forecasts in "two to three years" by using still-experimental numerical weather prediction techniques. Meteorologists would feed data into complex formulae and get out eight charts—computer-generated forecast maps representing "eight horizontal slices of the atmosphere, beginning at sea level and extending up to about 13,000 feet"—every 24 hours.[1] Everyone even remotely associated with the Meteorology Project, which was struggling to produce a working three-level model, must have read that statement with amazement. Still a dream in 1952, an eight-level model would start just above the surface and would have to top out at 30,000 feet to allow a 500-mb steering-level forecast. This disconcerting bit of media coverage notwithstanding, the race to operational numerical weather prediction was on.

While the Navy, the Air Force, and the Weather Bureau jockeyed for position, the Meteorology Project redoubled its efforts to ensure the availability of a viable operational model. It would leave decisions outside the project's purview—computer selection, data handling, and the adequacy of data coverage—to the weather services. In the meantime, project members continued to hope that numerical weather prediction would not be oversold to the public.

The Americans were not the only ones preparing to move from research to operations. In Stockholm, Carl-Gustav Rossby's group was also preparing to go operational. The Swedish team had fewer bureaucratic hoops to jump through—would they be the first to produce usable prognostic charts?

After 6 years of research and development, the Meteorology Project and its military and civilian supporters were ready to bring numerical weather prediction from a theoretical vision to an operational reality. The next 3 years would bring the project to a close, and a widespread operational and theoretical numerical weather prediction effort, with its great influence on twentieth-century atmospheric science, would begin. Governmental control over scientific endeavors, the use of the media to spread information on science and technology, and the importance of practical considerations over research when science moves from the theoretical to the operational realm, would all become important issues as the Joint Numerical Weather Prediction Unit took shape.

Theory Takes Aim at Operations

While the Weather Bureau considered how best to move numerical weather prediction from research project to operational practice, the Princeton team continued model development. Past modeling efforts had been more about atmospheric understanding than weather forecasting, but with the push to go operational, the Meteorology Project's models now needed to pass the usefulness test. As important as theoretical work was to developing numerical techniques and a robust atmospheric theory, gaining knowledge about the atmosphere was no longer sufficient—they needed realistic *predictions*. To see if their forecasts matched reality, the team members would need analyzed data to compare with model output. And for that, they would need theoretically savvy synopticians—synopticians who were not available in the United States. So once again the project looked toward Scandinavia. Project members also looked to the United Kingdom for additional help from dynamicists. And they prepared to bring weather service representatives—the men who would take numerical weather prediction operational—in for training. Under pressure to produce computer-generated weather forecast maps that met or exceeded the accuracy of hand-drawn maps, the Meteorology Project's already frenetic pace picked up.

Throughout 1952, the Meteorology Project concentrated on short-term forecasting goals as well as on more theoretical aspects of general atmospheric circulation. The latter would be especially important as project members attempted to extend the length of numerical forecasts. Forecasts also had to become more sophisticated, offering guidance for predicting cloudiness and precipitation. After all, when people wanted a forecast, they were interested in precipitation—would it rain, snow, hail, drizzle, or not? For numerical weather forecasting to be successful, the modelers had to produce significantly more than steering flow prognoses. They spent considerable time determining atmospheric flow changes for 24 and 48 hours—a necessary, although not sufficient, condition for predicting cloudiness and precipitation. They stuck with their original modeling plan of starting simple, analyzing shortcomings, and making small changes before running the next version.

Just as the group developed a simplified baroclinic model in mid-summer 1952, hardware failures began to have an adverse impact on model runs. From mid August through mid September, team members only got 11 hours of computer time. With daytime hours reserved for machine maintenance, project members were forced to run their models in the evening. The eagerly awaited magnetic drum—expected to help with memory deficiencies—would not be ready for another month or two. So the team, under Norman Phillips's guidance, worked on the general model using the minimum number of layers until the hardware situation improved. Model advances awaited additional memory and a computer that was operational more hours than not.

Sharing his frustrations with Jule Charney, then visiting the Astrophysical Institute in Oslo, Phillips reported that he had just managed to eke out a single 12-hour forecast on the barely operable computer and was sending the resulting charts to him in Norway for perusal. In Phillips's opinion, the finite 12-hour change was disappointing—only very slightly improved over the corresponding barotropic model run. However, he thought the initial tendencies looked quite good. Phillips did not yet understand the major error—the weakening of the height fall center—because the rise center, on the contrary, had "behaved well." Perhaps they had made a poor assumption or the mathematics was bad. Round-off error (due to rounding up or down to the next closest digit) was unlikely because the error was too large, and although truncation error (due to dropping off digits) could have caused it, Phillips could not understand why it would have been systematic.[2] They were making progress, but needed to do much more work.

Complicating the Meteorology Project's efforts, as always, were personnel shortages. Navy Lieutenant Albert Stickles, a recent graduate of the Naval Postgraduate School's meteorology program, had been assigned to the newly established Navy billet (position) with the project. Although Stickles would be attached to Princeton University's Naval Reserve Officer Training Corps unit for administrative purposes, his only duty would be as a project researcher.[3] The Navy's Daniel Rex assured Charney that Stickles had a personal interest in numerical weather prediction and was professionally competent in mathematical methods as they pertained to meteorology.[4] He would be the first of four weather service members assigned to the project for training, and would ultimately move to the Joint Numerical Weather Prediction Unit.

The project was still short of people—particularly with Charney overseas and Phillips, home alone, trying to lure Briton Eric Eady and Norwegians Arnt Eliassen and Ragnar Fjørtoft to Princeton for another tour of duty. Fjørtoft could be available as early as February 1953, but no later than September. Eady was available for 6 months starting in September. Eliassen would not be able to come until early 1954, but would then be available for 2 years.[5] With these additions, the project would have a more solid personnel situation in the future.

The lack of synopticians remained troublesome. Why was the Meteorology Project unable to find synoptic meteorologists within the United States who could do the job? Why did they have to be imported from across the Atlantic? The short answer is that synoptic meteorology was not held in high regard in the United States, and neither were its practitioners—unlike their foreign counterparts. The lack of interest in synoptic meteorology was not only a challenge for the project, which had a very small, albeit crucial need for suitably trained personnel. It was also a challenge for the largest employer of synopticians: the Weather Bureau. In a memorandum to some of his division heads marked "ADMINISTRATIVELY RESTRICTED," Francis Reichelderfer bemoaned the fact that so "few meteorologists really find their principal interest centered in the daily weather picture." He wanted some answers. Why were people being led away from synoptic meteorology?[6] One Weather Bureau man thought it boiled down to "you have it or you don't." There were those who experienced the poetry of weather and found "communion with the infinite." Others saw weather as contours on a piece of paper, and they did not become synopticians.[7] On a less poetic note, there was a financial reason: the bureau's synoptic meteorologists were not rewarded with pay grades equivalent to those of their more theoretical brethren. The analysis section's supervisor held a lower pay grade than those leading other sections. District forecasters could never advance beyond GS-11, a comparatively low rung on the US government's ladder of salaries for scientists. They were chronically undermanned, rarely selected for graduate study, and allotted no research time. Their French counterparts who were lead forecasters—i.e., those who led a forecast team of several people—worked one week on, one week off, and then had two weeks to do research. In marked contrast, the Weather Bureau's lead forecasters worked almost around the clock, seven days a week, and got recognition—the bad kind—only when they "busted" (that is, made a poor forecast). Until the situation changed, synoptics would hold no allure for meteorologists desiring career advancement.[8] Thus the Meteorology Project would have to depend on its Scandinavian contacts to fill the requirement, for it was the synopticians who were needed to move numerical weather prediction from development to operations.

The grim staffing situation, which had left only Charney, Phillips, and Smagorinsky working in Princeton, had gotten some relief during the summer of 1952. Ernest Hovmöller, who had worked on jet stream structures and aerological studies of cyclone structure with Rossby's Stockholm group, arrived in July to work on a synoptic investigation related to the transition from barotropic to baroclinic models. On a leave of absence from the Swedish Meteorological and Hydrological Institute (SMHI), he was obligated to return to Stockholm in January 1953. Von Neumann desperately needed Hovmöller to stay until at least September 1953 to complete this synoptic work. In fall 1952, von Neumann proposed to the SMHI's director, Anders Knut Ångström, that Hovmöller return to Princeton after a short trip back to Sweden in early 1953. In pressing his case, von Neumann wrote: "His association with us is perhaps the first example of the kind of cooperation that will ultimately have to take place between theoretical and synoptic meteorologists, if and when numerical forecasting is integrated into the governmental weather services."[9]

Von Neumann's argument was certainly valid. But Ångström rebuffed him, proposing that Hovmöller be replaced by fellow synoptic meteorologist Roy Berggren. Von Neumann was willing to take Berggren, but wanted Hovmöller, too, since their abilities would "complement each other admirably." Finances seemed to be a major stumbling block, but von Neumann was sure something mutually agreeable could be worked out. He suggested that Ångström wait to debrief Hovmöller in Stockholm before making a final decision.[10]

Back in Stockholm, Hovmöller provided a full account of the progress in Princeton to Ångström and to Rossby, who was pleased that project members were producing a significant number of computer forecasts. After Rossby had seen Ångström's latest letter to von Neumann about Hovmöller, Berggren stopped by to discuss the Princeton group with Rossby. Rossby then realized that Berggren had been "somewhat of a pawn in this peculiar chess game in which also Hovmöller is one of the pieces." Rossby told Berggren he could make a better contribution to the project if he reviewed related theoretical topics thoroughly before departing for Princeton. Berggren readily agreed with that plan.[11] However, von Neumann, desperately needing an outstanding synoptician, felt compelled to approach Ångström again. In a despairing note, he wrote: "To obtain such a person we are willing to bring him over from Europe and we agree with you that we should pay the costs, providing of course that he comes not as a student or trainee, but as an employee." Von Neumann was still paving the way for more collaboration with the SMHI so that its personnel could receive training in numerical methods, and, of course, so that the Meteorology Project could get its synoptic work done. Ångström, however, was unmoved. Hovmöller remained in Sweden. Berggren alone would join the Princeton team in March 1953, and would stay until the end of the year.[12]

The continuing saga of the synopticians was not the only item on the personnel agenda. Rossby's Stockholm group, awaiting delivery of the Swedish Air Force's BESK computer, needed to prepare by practicing on an operational machine. Thus, Bert Bolin announced in early 1953 that he wanted to get time on the new IAS computer as soon as possible. The only potential scheduling conflicts would occur in the summer. Rossby would be out of town, and if Bolin left too there would be no Swedes at the Meteorological Institute. So if Bolin could join the Princeton group in the spring, he might have to return to Sweden for the summer and then fly back to Princeton in the fall.[13]

::

Despite persistent personnel shortfalls, the Meteorology Project's pace accelerated once again in early 1953. With the IAS computer finally working well, the team was busy running and modifying models. The staffing situation was strengthened with the addition of new foreign members. Six full-time meteorologists were working with the group, including three new faces: Roy Berggren, Briton Andrew Gilchrist, and Kanzaburo Gambo of the Japanese Meteorological Agency (which was preparing to pursue its own numerical weather prediction program). Charney was delighted to hear that Bolin and Phillips would be trading places, a move that would allow for more cross-pollination between the Princeton and Stockholm groups. He hoped Bolin would be able to stay at least until the end of 1953, and longer if possible, since Fjørtoft and Eady would be in Princeton for collaboration. Charney was counting on Bolin to take Phillips's place as their "man-machine interface"—and to do so, they would need to overlap at the project.[14] Since he had to stay in Stockholm for the summer lest some "foreign person" get "stuck" with handling the Institute's affairs, Bolin suggested that he reach Princeton in August to train with Phillips. That would allow Phillips to arrive in Stockholm by fall, when Rossby's group would really need him for its computer work. Bolin told Charney that he planned to stay at least 6 months.[15] In a swift response, Charney informed Bolin that he (Charney) and Rossby had jointly decided that Bolin should start his work in Princeton on 1 September 1953. Bolin could remain in Stockholm through the early part of the summer, and could tour the United States with his wife in late summer.[16]

Not only had the regular staffing situation improved in early 1953; the four weather service trainees had arrived, and several European meteorologists would be stopping by for visits. Work had begun on the baroclinic model, which was in the coding stage by late March. Team members had also been able to speed up the barotropic model; the 24-hour forecast now took only

5–6 minutes of machine time—a dramatic improvement over the 24 hours it had taken on ENIAC only 2 years before. Team members also did another run on the November 1950 storm data using a significantly improved three-level baroclinic model for both 12-hour and 24-hour forecasts. Unlike the earlier runs, the model output did not deteriorate with time, and the storm prediction was better.[17] By June 1953, team members had run additional forecasts for the November 1950 storm as they further developed the baroclinic model. Instead of using 200-millibar (40,000 feet, 12,000 meters), 500-millibar (18,000 feet, 5,500 meters), and 850-millibar (5,000 feet, 1,500 meters) data, they used 400-millibar (23,000 feet, 7,000 meters), 700-millibar (10,000 feet, 3,000 meters), and 900-millibar (3,000 feet, 1,000 meters) data. The 900-millibar forecast accurately predicted the storm's rapid development—a very significant development. Overjoyed, the group reported that this was the "first successful attempt to forecast cyclogenesis by purely numerical methods." Team members had then started to make model improvements that would include previously neglected non-linear terms, using iterative processes in the solution of non-linear equations for the first time. Thus, the results would interest mathematicians as well as meteorologists.[18]

To advance the predictive models, team members also had to find an effective objective analysis program. The reason was simple. To obtain input data, charts for each atmospheric level had to be analyzed and gridded, and the data then had to be extrapolated to grid points. The grid-point values were then punched on cards or paper tape and fed into the computer. With an objective analysis program, the computer could perform all these steps, thus reducing subjective data interpretation and (it was hoped) the number of initial errors. The team members attempted a new approach by changing the radius surrounding the data so as to always provide for a minimum number of data points. Those data-point values would then be analyzed by the machine to arrive at one value for the grid point within the circle. From experience, team members knew this new method probably would not work the first time, but they planned to learn from each attempt and modify the method until they got the desired result.[19]

Project members were not likely to learn much more by continuing to run their models on the Thanksgiving 1950 storm data. They needed more case studies with which to compare their model output. Therefore, team members began analyzing maps for two additional cyclogenetic periods in the eastern United States.[20] After successfully running these data sets on the three-level model, they determined that large-scale mid-latitude storms were predictable, quasi-geostrophic, and quasi-isentropic—good news for operational applications that were now less than a year away.[21]

As they continued cleaning up their programs for operational use, the Princeton team also investigated aspects of the earth-atmosphere interface and atmospheric physics that had not been included in simpler two-level and three-level models—for example, the influence of mountain ranges, the vertical advection of vorticity, and the vertical propagation of energy. Project members also gained more experience with multi-level models by programming two five-level models (one for the IAS computer and one for an IBM machine). By March 1954, just a few months before the joint unit's opening day, they had run both programs out to 5 hours and were planning to extend that to 24 hours. They noticed that cyclogenesis had begun at the ground and worked its way up into the atmosphere in both models. In preparation for longer period predictions, the group members also wanted to explore how far into the future they could successfully extend the predictive period if they took into account energy sources and sinks, as well as the non-homogeneity of the earth's surface with respect to heat, water vapor, and momentum transfer. The models had to describe the essential processes governing the life cycle of a single large-scale atmospheric system and account for the general circulation of the atmosphere. From previous work, it appeared that it was necessary to have at least a three-level model to predict cyclogenesis. However, team members worried that they had not conclusively ruled out the efficacy of the two-level model. They had gotten good results from two-level models, and the results from the three-level models seemed to depend on the chosen levels. Additional investigations would be necessary.[22]

Although the research focus was heavily practical in the year before the joint unit became operational, the Meteorology Project's theoretical work was far from complete. There was still much to be discovered about atmospheric circulation and much work left to be done before longer-range forecasts (in excess of 36 hours) would be useful. Once their models were in the hands of the operational unit, they could turn back to more theoretical research, but as team members had been scrambling to debug and prepare their models for real-time operations, their weather service counterparts had been doing their part to ensure the new operational unit would be ready to use them.

The Weather Bureau Gets Started

Weather Bureau Chief Francis Reichelderfer, a strong proponent of theoretically influenced methods of weather forecasting since his days as a Navy aerologist, had aggressively encouraged John von Neumann to use his new computer to solve weather forecasting problems. Throughout the life of the Meteorology Project, Reichelderfer had periodically sent Harry Wexler, head of the Weather Bureau's Scientific Services Division, up to Princeton to check on its progress. He had also offered the assistance of bureau personnel in securing and analyzing data for the project's use. In fall 1952, as a member of the Joint Meteorological Committee of the Joint Chiefs of Staff, Reichelderfer had been in an excellent position to influence the direction of a joint operational unit manned and funded by all three weather services. Just as the joint analysis function of the Weather Bureau-Air Force-Navy (WBAN) analysis center had been centralized at the Weather Bureau, Reichelderfer had intended to do everything possible to ensure the bureau's position at the forefront of operational numerical weather prediction. To guarantee that he was not out-maneuvered by his military counterparts, Reichelderfer had started early to create an adequate support structure.

Reichelderfer selected Joseph Smagorinsky, still on leave from the Weather Bureau to complete his doctorate, to head a pre-operational numerical weather prediction unit: the Numerical Forecasting Group, whose purpose would be to indoctrinate bureau personnel in numerical techniques by having them do limited hand computations. Smagorinsky's group would also recommend changes to observation methods on the basis of anticipated data needs.[23]

Rejoining the bureau in January 1953, Smagorinsky found a lengthy list of high-priority tasks thoughtfully provided by Charney. Among these tasks were directing extensive hand calculations, determining large-scale weather elements from numerically predicted flow fields, and directing research on short-range and long-range predictions. Smagorinsky would have to do significant work on data acquisition and handling—determining the minimum amount of data for successful model runs, communications requirements for obtaining data and disseminating forecasts, and methods of electronically checking data and performing objective analyses. Furthermore, he would have to introduce the "philosophy, physical basis, and techniques" of numerical weather forecasting to the bureau.[24] There was no guarantee that numerical weather prediction would be an easy product to sell to beleaguered, marginally paid bureau forecasters who had limited professional training in meteorology—even assuming the numerical method lived up to its advertised potential. Accustomed to doing all analysis and forecasting by subjective techniques, forecasters were not likely to look with favor on predictions spitting out of a new-fangled computer. To ensure success, Smagorinsky not only had to be a masterly organizer of data, communications, and computers; he also had to be fully cognizant of model developments, and an excellent salesman.

Smagorinsky grappled with two immediate needs of his nascent Numerical Forecasting Group: people and equipment. He needed at least two full-time mathematics-savvy synopticians, two part-time statistical clerks, a full-time assistant, and a "simple electronic computer."[25] Personnel with both synoptic meteorology and mathematics skills were essential. Unfortunately, very few people possessing that combination of skills were willing to work for the Weather Bureau.

In addition to hiring personnel and efficiently handling data, the bureau needed to ensure an adequate stream of upper-air sounding data. This was an expensive proposition, since they not only needed *more* soundings, they needed *more widely distributed* soundings. The bureau barely had sufficient funds for their own upper-air stations, much less for upper-air stations that had been installed in developing countries during the war. Reichelderfer could not hide his impatience with military colleagues who thought funds would be found somewhere. They were unable, he groused, to recognize a bad fiscal situation when it presented itself. Reichelderfer asked his people to develop ways of getting needed upper-air information for considerably less money.[26]

By late spring 1953, personnel training and data-handling consumed Smagorinsky's time. Despite the bureau's budget limitations, he successfully argued that at least three of their meteorologists should be sent to George Platzman's new ten-week summer training program at the University of Chicago. The planned course, expected to cover the logic, physical basis, and techniques of numerical forecasting, would help the bureau maintain its perceived edge in operational numerical weather prediction.[27]

Smagorinsky recommended that the bureau pursue techniques and equipment for automatic data accumulation, handling, and transmission required for numerical weather prediction. Upon hearing this news, von Neumann and Charney were delighted. For once, there was hope of finding a way to fix 30 years' worth of "patch-work and improvisation" that had characterized the bureau's data handling. The typically overextended von Neumann even volunteered to be a consultant on this part of the project if it would speed up the process. Von Neumann and Charney were also enthusiastic about the bureau's efforts in objective analysis. Charney's team certainly did not have time to devote to it, and with the IAS computer inoperable for the next 2–6 months, computer time would not be available.[28]

By July 1953, the Weather Bureau had adopted Smagorinsky's suggestion to study Automatic Procurement and Processing of Data (APPOD)—ADP in more recent terminology. The extant data-handling situation was abysmal. When data came in via teletype, they were punched on paper tape. Then, instead of processing the data directly from punched tape, they were transferred to several other media. The original tape was then thrown away.

Data handling consisted of multiple steps and was not a single problem—it combined several problems, starting with instrument design and ending with forecast dissemination. Unless Smagorinsky and his small group addressed and fixed the possible error sources at each step, too much time would be devoted to processing observational data. And that would mean that a computer could generate a forecast in just a few minutes, but only after many hours had been expended trying to collect and feed in the data. Since speed was one of numerical weather prediction's main selling points, failure to fix data problems would eliminate much of its promise. Under consideration were ways of automatically retrieving surface and upper-air observation data, for instance readings of wind velocity, temperature, pressure, precipitation, cloud cover, radiation, and visibility. The Army Signal Corps had worked on automated observation instruments—initially designed for remote areas—during World War II. One idea for obtaining hard-to-get upper-air information from oceanic areas was to station automatic microwave relay stations at sea that would forward data transmitted from dropsondes (an instrument packet launched by rockets instead of carried up by balloons), but it was extremely expensive. If collected data could be transmitted and received by microwave, then written directly to magnetic tape, computer operators could greatly reduce processing time.[29]

Data verification was related to data handling. In the early 1950s, technicians evaluated data for accuracy. An incoming report that had taken a "hit," and was showing (for example) an unrealistic temperature or wind speed, would be tossed out before it adversely affected the model run. If a computer could evaluate raw data observations and could automatically eliminate spatially and temporally inconsistent data, then it could also be used to smooth out small-scale variations and allow for easier analysis and interpolation between grid points. This seemed a good plan, but sometimes the "odd" report *was the correct report*, which experienced forecasters could spot. If thrown out, the resulting model output could be badly skewed. Human evaluators were still needed, but if a machine could perform the initial data screening, it would certainly speed up the process.[30]

Once the computer had produced a chart, forecasters would need efficient ways of getting a hard copy. This part of the process had to be automated, perhaps with a mechanical plotting device. And once the plot had been made, the bureau would need to get the product out to forecasting stations—ashore and afloat. High-speed facsimile broadcasts would get the charts to local forecast centers, which could store them on a memory device such as a magnetic drum. The estimated price tag for Smagorinsky's staff of eight meteorologists and assorted technical specialists to do this work was $60,000.[31]

In mid 1952 the Weather Bureau had not been ready for operational numerical weather prediction, but it had asked the right questions and developed a plan to answer them. Smagorinsky had secured help with both training and data handling—neither of which had concerned the Meteorology Project. Now it was just a matter of making operational numerical weather prediction a reality.

Jointly Speaking

Having a meeting and deciding to "go joint" were easy tasks compared to the difficulties in implementing a joint operation. Such an undertaking required participants who could set aside their own personal agendas and concentrate on successfully melding people, institutional cultures, equipment spaces, and funding from different sources with minimal in-fighting. If the proposed joint operational center for numerical weather prediction were to become a reality, those interested in making it happen would need to move early and keep abreast of the situation. The Navy moved first.

In fall 1952, Commander Daniel Rex of the Office of the Chief of Naval Operations had announced the Navy's intention to negotiate a joint agreement among the Navy, the Air Force, and the Weather Bureau concerning the organization, the scope, and the objective of a "national numerical forecasting (or computing) center."[32] By May 1953, after an impromptu meeting in Princeton involving Smagorinsky and Colonel George F. Taylor of the Air Weather Service, the Air Force and Weather Bureau agreed that all of the weather services would need to stick together to meet labor and financial requirements surrounding operational numerical weather prediction.[33] The Weather Bureau made this very argument in a point paper written for the Joint Meteorological Committee of the Joint Chiefs of Staff in June, and dropped a disconcerting bit of news: they had received reports that Sweden, England, and West Germany were planning their own forays into operational numerical weather prediction by January 1954.[34]

The possibility of being overshadowed and outperformed by European groups undoubtedly provided some impetus to establishing a joint operational unit. By mid 1953, all of the US weather services were actively supporting some kind of joint operational approach to numerical weather prediction. The next step: the establishment of a series of ad hoc committees to provide direction.

Within two weeks of receiving the Weather Bureau's point paper, the Joint Meteorological Committee (JMC)—originally established during the early years of World War II to address the provision of meteorological support during national emergencies and still in active operation—created the Ad

Hoc Committee on Numerical Weather Prediction. Composed of representatives from each weather service, and chaired by Daniel Rex, the committee acted quickly to move operational numerical weather prediction forward. Two months later, these committee members, after consulting with von Neumann, Charney, and others from the IAS projects, devised a detailed plan for creating an operational numerical weather prediction unit by 1 July 1954.[35]

The ad hoc committee set four goals to be met before the unit's opening day: model development, computer acquisition, personnel training, and a facility to house the people and the computer. The first goal had been met by the Meteorology Project. Numerical analyses and prognoses had shown sufficient skill (based on placement of high-pressure and low-pressure systems) to make them competitive with the best subjective methods. That was a very optimistic conclusion, but one that had to be made to keep the project viable. (Had it concluded that the models could not compete with subjective methods, the proposal to establish the joint unit would have quickly died.)

To meet the second goal, the weather services would have to secure a sufficiently powerful computer. In 1953, few computers could handle the meteorology models, and they were not available off the shelf. The International Business Machines Corporation had produced a Type 701 electronic computer—closely modeled after von Neumann's machine—which could be used for meteorological work. IBM could have a leased version of this machine ready by 1 October 1954.

Fortunately, the third goal—sufficient trained personnel—already had been met: the three weather services, at last, could identify enough meteorologists and meteorological analysts to serve with the proposed unit. In addition, the last goal—a place for the unit to call home—appeared secure based on information provided by the JMC itself. The JMC advised the Ad Hoc Committee that space would be available near the WBAN Analysis Center—which had been in the decrepit Weather Bureau headquarters building, but was preparing to move to new spaces in Suitland, Maryland, in spring 1954.[36]

While everyone concerned agreed that it was best to "go joint," in reality military and civilian organizations operated (and continue to operate) differently. Military personnel were reassigned frequently, and high personnel turnover would be a disadvantage. Therefore, the military weather services were encouraged to extend personnel "tours" (i.e., assignments) for as long as possible.[37] This was very important. Most military assignments were only 2 years long—some were even shorter. Without tour extensions, military personnel would leave for new assignments just about the time they became productive team members.

With the time ripe to form an unprecedented operational unit, the committee wanted to ensure that the unit's organizational structure would be flexible enough to quickly incorporate new research results. The unit's forecasts would need to be examined closely and carefully, and the models tweaked to effect small, steady improvement. It would be up to the weather services to ensure the ultimate success of operational numerical weather prediction.

The committee recommended that work on the joint unit move forward. Although the Weather Bureau would assume administrative responsibility, all three weather services would provide funding. Having fulfilled their purpose, members of the Ad Hoc Committee proposed establishing a new steering committee responsible for selecting the unit's director and helping him implement the plan. They also decided on the mission of the Joint Numerical Weather Prediction Unit (JNWPU):

To produce on a current, routine, operational basis, prognostic charts of the 3-dimensional distribution of relevant meteorological elements by using numerical weather prediction (NWP) techniques, in order to improve the meteorological forecasting capabilities of the participating weather services.[38]

Operationally, the unit would undertake data analysis and processing that were beyond the capabilities of the WBAN Analysis Center. It would compute prognostic charts and create products of greatest benefit to field forecasters. Additionally, the unit would verify computer-generated products—monitoring quality and suggesting model changes for further improvement. It would develop objective analysis methods and improve data-handling techniques, extend models geographically, and adapt models for longer forecast periods. The unit would liaise with other organizations, particularly those conducting numerical weather prediction research, and determine the applicability of new research results to operational models. It would also conduct in-house training of personnel to maintain optimal personnel proficiency.[39]

Since the unit was starting from almost nothing, the committee anticipated a three-month shakedown period during which the unit would prepare and distribute one set of prognostic charts daily. Unit members would focus on developing and standardizing an operational routine. By placing the JNWPU adjacent to the WBAN Analysis Center, the services hoped to eliminate duplication of effort—a long-time concern. In the past, concerns over apparent duplication of effort had led to retrenchment in military meteorological organizations at the end of national emergencies. The Weather Bureau did not want its efficiency questioned yet again on this high-visibility, high-cost project with still uncertain results. The WBAN could provide the unit with plotted maps and additional data as needed. Although the unit might need to perform its own analyses to meet specific numerical prediction requirements, i.e., those needed to provide initial grid-point values, it would use WBAN analyses when possible.[40]

The committee set forth tentative goals and operational procedures for the JNWPU. At first, unit members would use a simple atmospheric model to produce prognostic charts of pressure surfaces, vertical velocity, average temperature, and perhaps large-scale precipitation rates. The weather services could then choose which products to transmit via their facsimile broadcasts. The meteorologists would also evaluate the computer-created prognoses and make iterative improvements to the model. However, it would not be the unit's task to develop new models—those would come from research and development sites and then be adapted for operational applications. Unit members would be able to extend the size of the geographical area and the forecast period. They would consider data processing improvements separately, with much of the emphasis on objective analysis techniques.[41]

Anticipating potential inter-agency coordination battles, the committee recommended that the JMC appoint a steering committee composed of service representatives to hear conflicts over service personnel, requirements being placed on the unit, and technical matters external to the unit. For example, each weather service would have unique mission requirements demanding particular products. If the unit tried to meet too many of these service demands, it could find itself unable to complete its primary mission. This committee would sort out such conflicting requests. The director of the JNWPU would report to the Weather Bureau's chief on all administrative matters including finance, civilian personnel, and logistical support. A scientific advisory group composed of subject matter experts, including scholars from meteorology, electrical engineering, and mathematics, would visit periodically and provide technical advice.[42]

::

Having set forth the nascent unit's mission and a general outline of its duties, the Joint Meteorological Committee disbanded the Ad Hoc Committee on Numerical Weather Prediction and reconstituted the members into the Ad Hoc Committee for Establishment of a Joint Numerical Weather Prediction Unit to help navigate the bureaucratic maze that awaited. As the unit's needs changed, so did the committees. But two impediments to progress persisted: equipment and personnel.

In the summer of 1953, the first ad hoc committee had concluded that the IBM 701 "Defense Calculator" was the best computing machine available. Designed to meet the demands of the Defense Department and the aerospace industry (which in fact used almost all of the 19 extant machines), this machine's design, logic, and high-speed memory were almost identical to those of the IAS computer.[43] With an extremely flexible input/output scheme, and with IBM's promise to cooperate on automatic data processing developments, the 701 was clearly a good choice. Indeed, committee member did not discuss any other options, and they may not have been aware that one existed. They did know that if they submitted a letter of intent to IBM before 30 September 1953, the computer could be available by 1 October 1954—a full year later, and 3 months after the JNWPU was to open its doors. The amount budgeted for the computer was $200,000.[44]

For the second ad hoc committee, selecting a computer and providing an appropriately engineered space to house it became its two main technical challenges. Committee members had three options for obtaining the required computing machine: order it custom-built, purchase it off the shelf, or lease it. A custom-built machine would be prohibitively expensive and limit their flexibility to upgrade to a faster computer with more memory when one appeared. A computer purchased off the shelf would similarly limit future choices. Therefore, leasing seemed the best option. The unit could make equipment upgrades without investing large sums of money. Just as important, the leasing company would provide all the maintenance—a significant savings in manpower costs and hassles.[45]

The committee would have to address the 701's siting requirements well before delivery. Spaces adjacent to the WBAN had sufficient heating, lighting, and power systems, but not the required 30 tons of air conditioning. The Weather Bureau would need to raise the floor to allow for cabling and air conditioning ducts. IBM engineers would also need an office and storage space for spare parts and equipment[46]

The committee knew what it wanted, and the weather services knew what they needed, but those wants and needs soon crashed head-on into the reality of the fiscally conservative Eisenhower administration. Reichelderfer received a query from Assistant Secretary of Commerce James C. Worthy as to why the JNWPU needed a dedicated computer. Why not share the Census Bureau's Machine Tabulation Facilities? If the Weather Bureau did plan to use the Census Bureau's machine, Worthy wanted a detailed description of the "nature and scope" of the proposed use.[47] Reichelderfer was distressed by this possibility. Attempts to save money by sharing computing equipment would be untenable if the unit were expected to run its models only when other agencies were not using their computers.

Wexler took on the task of responding to Worthy's concerns, forcefully arguing that Census Bureau's facilities were wholly inappropriate for the joint unit or the extended forecast division. The Weather Bureau not only needed a more sophisticated computer than the Census Bureau owned, it needed to use it 70 hours a week on a very strict schedule. During the first year, the run time might double to 140 hours per week. Even during the shakedown phase, staff members could not stand around waiting their turn behind Census Bureau personnel. If computer sharing were forced on them, the Weather Bureau would face two unhappy options: moving the extended forecasting section to the Census Bureau, or hauling the data on punched cards to the Census Bureau. Neither option was tenable. Separating the extended forecast meteorologists from the rest of the bureau's professional staff was nonsensical. Likewise, driving numerous large decks of cards around the greater Washington area during rush hour, nasty weather, or other traffic disasters would adversely affect the creation of timely meteorological products. Worse yet, what if someone dropped the deck of punched cards? The Weather Bureau, Wexler wrote Reichelderfer, required a dedicated computer.[48]

However, Reichelderfer demanded a more politically astute response to Worthy's question, so Wexler tried again a week later. Sensitive to Reichelderfer's fiscal worries, he argued that leasing a computer would be not only economically prudent but also technologically smart. Extant computing power was only marginally able to meet the challenge of running realistic atmospheric models and processing data automatically. Newer models would soon outstrip the capability of any machine they chose, so being able to upgrade was a primary consideration. Perhaps when (in Wexler's words) the "situation stabilized"—when computer design slowed so that new, faster models were not continuously being made available—it might make more sense to purchase one. Now was not the time.[49]

While agreeing with Wexler's assessment, Reichelderfer nevertheless argued that they had to be attuned to their federal overlords: congressional authorities, the Bureau of the Budget, and high-ranking members of the federal government's executive branch who were convinced that too much money was being spent on "machine tabulation equipment." All were pressing agency heads to share computing equipment whenever possible. Under no circumstances did Reichelderfer want to appear uncooperative by insisting that his organization required a dedicated computer. Therefore, he proposed that the Weather Bureau "not give the appearance of obstructing the plan in the beginning by starting off with reasons why we cannot do it." He seemed to think that the reasons would "speak for themselves" once the Census Bureau and the Commerce Department reviewed the new unit's computing requirements.[50] It was risky for Reichelderfer to count on other agencies to see the wisdom of his thinking. Those agencies owned the machines—the Weather Bureau was coming with hat in hand. If Census and Standards thought the Weather Bureau would take over their machines, then he might count on their support as well for a dedicated weather computer.

Apparently the Weather Bureau's response had the desired effect, because computer arrangements initially moved forward. But less than a month later, another bureaucratic hurdle popped up before the intrepid and increasingly beleaguered operational numerical weather prediction pioneers. The Joint Meteorological Committee decided against leasing the IBM 701 without a competitive bid. And so in September 1953, the latest Ad Hoc Group started its quest for a competitive bid in an era of few potential bidders. Smagorinsky invited the only two firms with competitive machines—IBM and Remington Rand—to perform preliminary tests demonstrating their machines' capabilities. Von Neumann, Charney, Platzman, and several others from the Meteorology Project elected to help analyze the results.[51]

Meeting with company representatives in October 1953, the group asked them to run the three-dimensional quasi-linear model. IBM had a 701 available for the test, but declined to do the coding without compensation. Remington Rand was willing to do the coding without charge, but did not have a computer (ERA 1103) available to run it. IBM, as noted, could build and deliver a 701 within a year of receiving a letter of intent. Remington Rand's representatives had no idea when they could deliver an operational computer, or how they would handle maintenance.[52]

Designated as fact finders, Smagorinsky and IAS mathematician Herman Goldstine worked to assess the computers' suitability.[53] Within a couple of weeks, Remington Rand had found machine time and IBM had identified a program it could run on its 701.[54] Testing would start in December and be completed by mid January 1954.[55] Once test results were available, the Ad Hoc Group and its technical advisors would meet to discuss the outcome. They hoped to make a final decision by early February.[56]

Goldstine and Smagorinsky filed their report at the end of January, outlining how the complexity of the atmospheric models and the method of handling data affect run times. They argued that there was no point in selecting a computer that could not run the initial (and least complex) model and easily manage incoming data. Nor did it make sense to choose a machine that could handle the initial models, but fail to run models incorporating larger geographical areas, additional variables, or increased forecast periods. Smagorinsky and Goldstine anticipated that the first year's models would include large-scale atmospheric motions, assume an adiabatic (non-heat exchanging) and frictionless atmosphere, and consider an irregular lower boundary. In mathematical terms, the model represented an initial-value problem wherein the geometric boundary conditions were specified at all times. It would have three internal vertical grid points, be quasi-linearized, and have a lower-level boundary, i.e., it would not consider topography. The more general version of this model would have five to seven vertical grid points and an irregular lower boundary, and it would require solution of the three-dimensional Poisson equation. Smagorinsky and Goldstine estimated that the run time for this more sophisticated model would be at least five times greater than for the simpler quasi-linear, three-level model.

Other factors would also increase run times. Automatic data processing alone would require inverting nearly a thousand 10×10 matrices and use a significant amount of machine time. Increasing geographic coverage, and including moisture distributions for precipitation forecasts and three-dimensional trajectories for condensation computations in later models, would lead to longer run times too. Given these possibilities, Smagorinsky and Goldstine thought it likely that within a few years the required machine time would be an order of magnitude greater than for the test problem run by IBM and Remington Rand. Since future operational predictions would require a machine with a faster processor, the JNWPU's ability to function effectively would depend on the availability of more sophisticated computers.[57]

As it turned out, the IBM and Remington Rand machines produced solutions—complete with printed output—in the same amount of time. The IBM had a faster processor, but slower input/output devices; the Remington Rand was slower, but had a significantly faster printer. Since the bids were essentially the same, and IBM had a better maintenance record, the group recommended—once again—leasing the IBM 701 Defense Calculator.[58]

About the time Reichelderfer managed to convince the Commerce Department that the JNWPU absolutely had to have its own computer, a new question arrived from Under Secretary of Transportation Robert B. Murray Jr.: Why not just use the best parts of the IBM and Remington Rand computers to build a computer better than either one of them was individually? Once again, Wexler was dispatched to research the answer, while Reichelderfer patiently told Murray that it made more sense to lease a machine than to purchase one that would be obsolete in 2–3 years. Time was getting short. They needed to order the computer right away.[59]

Having once again tamed bureaucratic nightmares, the Weather Bureau's financial maven, Robert Culnan, negotiated the final details with IBM and sent the letter of intent.[60] Only 4 months remained until the JNWPU would open for business. Then, in late May, just when computing equipment procurement appeared settled, IBM announced a new computer—the 704.

Did the JNWPU want it? After tossing this idea around, the Ad Hoc Group decided to stick with the 701. They were already behind schedule, the 704 had no operational track record, and adopting it would delay delivery by several more months and would increase the cost.[61] The official delivery date for the 701 was now 1 March 1955. The men of the JNWPU had at most 10 months to test and refine their initial model.[62]

Almost a year had elapsed since the Joint Meteorological Committee had directed the establishment of the Joint Numerical Weather Prediction Unit, and the Weather Bureau had just signed a letter of intent for computing equipment. While the members of the Ad Hoc Group, their technical advisors, and the Weather Bureau's leaders had been addressing hardware acquisition, some of these same people had also been working on an equally important job: hiring.

::

With the Joint Meteorological Committee's decision to move ahead with the establishment of a new joint unit, the Ad Hoc Group needed to figure out how best to hire personnel for a tri-service organization. Mixing military and civilian personnel in the same organization historically has never been an easy process: doing it successfully requires the right combination of person-alities, a large dose of patience, and an outstanding leader.

The director, the Ad Hoc Group decided, should have a broad background in both synoptic and theoretical meteorology in addition to experience using mathematical and physical techniques in weather prediction. He would also need to be familiar with basic programming techniques. The members unanimously recommended Air Weather Service meteorologist George P. Cressman, who had earned a PhD from the University of Chicago during Rossby's tenure. A well-recognized authority on synoptic meteorol-ogy, Cressman had had experience with all three of the weather services that would constitute the joint unit. When informally approached by the group in September 1953, he agreed to fill the position.[63] This biggest challenge still remained: Cressman had only 9 months to find and hire his professional and technical staff.

Whereas Cressman would be primarily concerned with the unit's scientific work, the new assistant director, still to be appointed, would handle admin-istrative matters. This individual would have to be conversant with numeri-cal weather prediction but would not have to be a meteorologist. However, he would have to be *tactful*—and that is an understatement. Overcoming the rivalries among three competing weather services would require not only tact but also a huge amount of patience.

The remaining personnel would have to have meteorological and/or mathematics backgrounds. At least some of the new hires would come from the ranks of those training with the Princeton or Stockholm groups, or those working with Philip Thompson's Cambridge team. The four research and development meteorologists would collectively, although not individually, have to possess strong dynamic meteorology and mathematics abilities and be familiar with machine computations. They would also have to have extensive synoptic experience and knowledge of advanced prognostic techniques, plus "proven abilities to carry out independent developmental research." Assisting them would be ten meteorological analysts (five for each shift) with strong synoptic backgrounds and enough general knowledge of dynamics and numerical weather prediction to be useful team members. Sub-professionals, including those who would check and plot data, would round out the unit's meteorology component. The incoming mathematician would need extensive experience in numerical analysis and the programming of complex physical programs. He would be joined by three programmer-coders who would also be strong mathematicians with programming experience. They would be assisted by computer operators, who would be responsible for running the models.[64]

To forestall inter-agency personnel conflicts, the Ad Hoc Group's representatives worked closely together from the outset to get the joint unit off to a good start with capable, enthusiastic personnel. When it turned out that certain weather services could provide more people than others, the group decided to trade people for cash. The Navy, for example, had few people to assign to the unit, but was able to provide more funding. That the unit's proposed internal structure remained intact as hiring continued was nothing short of amazing. The JNWPU faced two immense challenges: it was an entirely untested organization creating untested meteorological products, and the personnel it needed were likewise entering an entirely new field that most of them were learning on the job. With the exception of difficulties in locating a suitable programmer-mathematician, staff recruitment went smoothly. Cressman, however, wisely decided to postpone filling sub-professional positions until his new organization had found a permanent home.[65]

By the time the Ad Hoc Group made its final report, and by the time the joint unit became a reality (1 July 1954), all but three professional positions had been filled. Of the professional core, seven each were from the Weather Bureau and the Air Force; three were from the Navy. The JNWPU was still short one meteorologist and two operators, but interviews were underway. The Air Force provided three sub-professional staff members and the Weather Bureau one. The remaining ten positions would be assigned from Weather Bureau assets.[66] As nightmarish as the hiring process could have been for the new JNWPU, surviving documents indicate that it went more smoothly than the acquisition of a computer and its associated peripherals.

Since the Weather Bureau would have administrative authority over the proposed joint unit, in summer 1953 Harry Wexler arranged the appointment of a financial representative. Robert Culnan became the unit's financial coordinator and established contacts with Air Force and Navy representatives. However, none of the weather services would be allowed to transfer funds until the Ad Hoc Group had ascertained the extent of space modification expenses, and that depended on the unit's exact location, for which a decision was not anticipated until early 1954.[67] The first of the year brought no decision, however, and without a firm location, the bureau could not develop a final budget. Despite that, the group decided that each service should transfer funds—about $31,500—to cover one-third of the proposed start-up cost to the Weather Bureau, and that Cressman should be authorized to expend those funds.[68]

By mid-summer 1954—*after* the establishment of the joint unit—space was allocated in Federal Office Building No. 4 in Suitland, Maryland, but not next to the WBAN Analysis Center (also in Suitland) as previously planned. However, this new location was adjacent to the future home of the new National Weather Analysis Center—the WBAN's successor. In the meantime, the JNWPU would occupy space made available by the Weather Bureau. The space assignment allowed budget planning to continue.

Of the original $94,500 in start-up funds, approximately $82,000 was by now earmarked to modify the spaces, install power and electrical fixtures, and pay for engineering services. The remaining money would be used for miscellaneous equipment and furniture. The total estimated expenditure for fiscal year 1955 was about $300,000, or a little over $100,000 per service. Since the Navy was providing fewer people, as noted above, its cash contribution was almost double that of the other two services. By providing more than one-third of the personnel, the Air Force reduced its expected fund transfer.[69]

While the three weather services had been kept busy shuffling people, money, and spaces to meet the opening day of the Joint Numerical Weather Prediction Unit, and while the modeling team in Princeton had been busy cleaning up analysis and forecasting models for their operational brethren, the Europeans had been making their own move to operational numerical weather prediction.

The Europeans Close the Gap

As noted above, the move to operational numerical weather prediction was not restricted to the United States. During the year preceding the opening of the JNWPU, other centers of activity were gaining ground in Europe and pushing toward their own forecasting units. Besides the Swedish team, the German meteorologist Karl Hinkelmann had signed a contract with the US Air Force for numerical prediction, "including the building of a machine"—an outgrowth of Thompson's earlier European visit. The US Air Weather Service wanted numerical weather prediction support for its assets in Europe, but was not able to provide them from the United States—computers had neither the capacity nor the speed to process all the data. The Air Force's solution was to establish numerical weather prediction centers, not unlike the JNWPU. Two members of Hinkelmann's team were in Stockholm, and Hinkelmann was scheduled to join the Stockholm group in January 1954.[70] So despite the Air Force's desire to participate in the joint unit, the Air Weather Service was working independently to expand numerical weather prediction to Europe, in addition to maintaining Thompson's group in Cambridge.

In late October and early November 1953, Philip Thompson made another European tour to assess the progress of numerical weather prediction in Sweden, West Germany, and the United Kingdom. After returning to the United States, he reported that the Europeans lagged by about 6 months in basic theory and 1–2 years in operational applications, owing to personnel shortages, lack of training, and non-availability of dedicated computers for numerical weather prediction. The Deutscher Wetterdienst was working on putting numerical weather prediction into operation, but would probably not do so before early 1956. The British Meteorological Office intended to add numerical techniques to its forecasting practice, but without a computer its efforts would be limited. Rossby's group "professed to have no definite plans for operational applications, but [would] have the capabilities for putting numerical methods into practice by early 1955." Since the Swedish team in fact began producing operational forecasts in 1954, it appears that Thompson was somewhat led astray by what he heard in Sweden. Thompson was authorized to offer Rossby the possibility of an Air Force research contract, which he was glad to take. But Rossby reminded Thompson that, because Sweden was a neutral country, the funds would have to be "decontaminated" via a civilian institution, perhaps the Woods Hole Oceanographic Institution.[71]

By late 1953, Rossby's Stockholm group was very busy on BESK, which Phillips reported was running quite well. Although output was slow (because of the printer), BESK was faster than the IAS machine on which it was modeled. When their magnetic drum arrived, the Stockholm team planned to increase grid size, the number of model layers, and forecast length.[72]

In early spring 1954, Smagorinsky went to Europe and reported that the British and the Swedes anticipated making daily operational predictions within 6 months.[73] It happened sooner than that. In mid June, Rossby informed Charney that the Stockholm team had made 23 barotropic forecasts for the eastern Atlantic and northern Europe, including two operational ones, on BESK. Having gotten good results, they were preparing to make operational 48-hour forecasts.[74] In contrast, the JNWPU's computer would not be available for at least another 6 months.

The Media and Numerical Weather Prediction

From the earliest days of John von Neumann's Computer Project, outlandish descriptions of what a computer could do for meteorology and weather forecasting had appeared in the media. As discussed in chapter 4, early press coverage in the *New York Times* linked numerical weather prediction and future weather control—links that were largely missing from stories by the early to mid 1950s. By then, weather control had become a press topic in its own right owing to Nobel Laureate Irving Langmuir's cloud seeding experiments. However, press reports exaggerating computer forecast capabilities made von Neumann, Charney, and others who thought the entire project was being oversold very uneasy. Indeed, project members were having difficulty persuading their fellow meteorologists to take their work seriously. Accounts of long-range computer-predicted weather in a few minutes of machine time were not helpful. One way to counteract fantastic press reports would be to meet the press. And so they did, even if their own stories were inconsistent.

In May 1953, Joseph Smagorinsky talked about numerical weather prediction while appearing on "Science Forum" on WGY, the General Electric Company's radio station in Schenectady, New York. Reviewing why people wanted to know the weather in advance, Smagorinsky explained that meteorologists had significantly increased their understanding of atmospheric physics since the turn of the century. He described how physical laws could be described by differential equations (equations that related small spatial variations with small intervals of time), which could be solved by electronic computers. Whereas meteorologists in the past had solved these systems of equations by hand, with computers integrating the equations they could expect more accurate forecasts. Smagorinsky continued: "The vision of Professor John von Neumann of the Institute for Advanced Study at Princeton, New Jersey made it possible to apply high-speed computer methods to the weather forecast problem—which has come to be known as numerical weather prediction." He also credited Rossby (as instrumental to the study of earlier meteorological work, which allowed current researchers to avoid the "pitfalls" of the past) and Charney (as having simplified the meteorological equations in a "rational" manner that bypassed "years of experimentation and research").[75]

Smagorinsky further explained that, to obtain forecasts by numerical methods, meteorologists needed wind, pressure, temperature, and humidity observations over large geographical areas, up to an altitude of 70,000 feet above sea level, and not more than 200 miles apart. A 24-hour forecast for New York City would require large amounts of data in a 600-mile radius around the city. For a longer forecast period or a larger forecast area, meteorologists needed a larger radius. Thus, to forecast for the United States, data were needed from the Pacific to the Atlantic and from the Canadian Arctic down to Mexico. Once the data were available, the forecast would take only minutes.

The public's concern that automation would eliminate jobs must have been on Smagorinsky's mind as he was preparing his presentation. His radio talk was primarily about numerical weather prediction methods, and yet he went on to explain that they would not result in the mass unemployment of meteorologists. Rather, meteorologists would just have more time to make sense of the atmosphere and thereby provide better, smaller-scale forecasts. Smagorinsky also admitted that meteorologists still did not understand many things (e.g., the sun's radiation, atmospheric turbulence, and pre-conditions for precipitation) that affected the weather. Until those factors were addressed, meteorologists could not attempt long-term forecasts of up to a year in advance. However, Smagorinsky assured his radio listeners that the Weather Bureau was addressing all these concerns so that it could continue to provide the very best weather information to the nation.[76]

Surely Smagorinsky was seeking positive press coverage that would enhance the Weather Bureau's reputation. Reporting back to his bosses, Smagorinsky wrote that his talk had received a good review from the *Schenectady Gazette*. He pointed out that it "once again indicates that publicity originating from the Weather Bureau can help us to have sympathetic relations with the press and the public." This was in contrast to a May 1953 *Fortune* article, "Tomorrow's Weather," that became a public-relations disaster. The *Fortune* article, which considered both weather modification and extended weather prediction, implied that the Weather Bureau was so conservative that it had refused to use ground-based seeding methods perfected by commercial seeders in its own trials. Although not overly reproachful of the bureau, it did put those working in the private sector in a better light. "It would seem," Smagorinsky continued, "that the Weather Bureau should seize upon every opportunity to educate the public (in a dignified manner, of course) on our efforts toward carrying out our primary mission." After Smagorinsky's letter reached Weather Bureau headquarters, someone—probably Harry Wexler—scribbled in the margin "Do we *not*?!!"[77] Apparently, the Weather Bureau was feeling beleaguered by adverse press coverage.

In October 1953, as the JNWPU project gained momentum, Science Service staff writer Ann Ewing attended an Ad Hoc Group meeting to get help with an article she had written about numerical weather prediction. Ewing's article described how a "giant electronic 'brain'" (computers in this period were almost always "electronic brains") would be making daily wind predictions that would then be used for local weather forecasts on an *experimental* basis within a year's time. Billing the undertaking as an experimental program run by the three weather services, the article went on to say that this "revolutionary method" was first developed at the Institute for Advanced Study. In fact the field was so new, Ewing declared, that few subject experts existed worldwide. After describing input and output, she asserted that these "brains" would "eventually eliminate most of the forecaster's *personal opinions* from his predictions." Although the formulas had not been entirely worked out, Ewing wrote that eventually meteorologists hoped to include a variety of energy sources (e.g., radiation from the sun and heat released during condensation, and energy sinks such as those due to evaporation) in their models. These improvements would allow forecasts for 5 days, 30 days, or even more. Long-range predictions were not expected soon.[78] Whether Ewing considered 5 days "long range" is unclear, but long-range predictions certainly were not going to appear any time soon. The spurious idea that forecaster's "opinions" would ultimately be eliminated from weather prediction was another common theme. Statements like these led the public to think that the forecasts they heard on the radio or read in the newspaper soon would come from computers.

Other publicity efforts presented the project's history differently. In November, the Office of Naval Research, with the approval of the Defense Department's Office of Public Information, issued its own press release in an attempt to gain positive attention for its role in numerical weather prediction. In that release, titled "Electronic Weather Forecasting," the Navy claimed credit for numerical weather prediction, accurately saying that the newly forming joint unit was an outgrowth of research "initiated in 1946 when the Office of Naval Research contracted with the Institute of Advanced Study in Princeton to study numerical prediction technique." The Navy went on to claim that the Meteorology Project had been established by the Office of Naval Research, which had given the task of developing the technique to von Neumann and Charney. This statement was not entirely accurate, since Charney did not come on the scene until 2 years into the project. After briefly describing the roles of von Neumann, Goldstine, and Bigelow in the design and development of the IAS computer, the press release asserted that "in January 1949 the Office of Naval Research *invited* the Geophysical Research Division of the Air Force to participate in the electronic forecasting technique," thus allowing the project to move forward even more rapidly than before. There was also a lengthy quote from "Dr." (instead of Commander) Daniel F. Rex of the Office of Naval Aerology, who compared the revolution in numerical techniques to the one spawned by Bjerknes's "Norwegian Wave Cyclone Theory." Numerical methods, Rex further asserted, would enable local forecasters to spend more time on the details of their local weather, since they would not have to draw their own prognostic charts.[79] Rex was certainly a PhD meteorologist, but it appears that the Navy, by referring to him as "Dr." instead of as an active-duty officer, was trying to attach more credibility to his statement. And in fact no available archival evidence indicates that the Navy asked the Air Force to join in sponsoring this project.

The flurry of press releases and articles did not escape the notice of the Ad Hoc Group (of which Daniel Rex was a member). Annoyed, the members of that group decided that whenever a significant development occurred, or when new information was available, they would prepare a joint statement and send it to von Neumann, Reichelderfer, Charney, Thompson, and the ONR's representative for their own use or for further distribution within their organizations.[80]

In early December, Wexler visited von Neumann and Charney in Princeton to discuss some of the recent press accounts. Von Neumann was upset that some of the articles had tended to place the greatest emphasis on the machinery instead of on the intellectual achievement that allowed the modeling to take place. Wexler subsequently advised Reichelderfer as follows: "We shall have to be even more careful in the future in cautioning reporters to avoid some of the objectionable features."[81]

But Wexler could play the press game too. In February 1954, he presented a proposed press release about the computer simulation of the "busted" East Coast snowstorm of January 1950. In the article, the data were run through one of Princeton's models, which successfully identified the storm and steered it in the correct direction. The implication was that numerical methods yield better forecasts. In the discussion that followed, the Ad Hoc Group expressed concern that the press was considering the results of the competitive tests between the IBM and Remington Rand computers as a test of numerical weather prediction techniques. Therefore, the success of those runs would be equal to the success of operational numerical weather prediction in readers' minds. The group wanted to make sure the press did not leave the public with the impression that numerical weather prediction already was a proven technique.[82] But there is no doubt that the word was out. Even the Boy Scouts of America's Editorial Service sent a letter to von Neumann asking about the work of the Princeton teams.[83]

Those developing numerical weather prediction certainly needed the help of the media—radio, newspapers, and magazines—to tell their story. Everyone involved desperately wanted the public to get these messages: numerical weather prediction was worth the investment of time and money, would lead to more accurate forecasts, and was the future of modern meteorology. But while the project's participants wanted the public and the scientific community to be sold on numerical weather prediction, they did not want them to be oversold. And although articles about "giant brains" forecasting the weather certainly attracted attention, there was always a nagging concern, particularly at the Weather Bureau, that if the result did not live up to the hype, what little credibility the meteorological community had gained during and after World War II would be diminished significantly.

The Joint Unit Comes to Life

On 1 July 1954, after more than a year of planning and negotiation, the Joint Numerical Weather Prediction Unit became a reality—a non-operational one initially, because it had no computer and would not for at least 6 months. However, its personnel still had plenty of work to accomplish. George Cressman laid down four primary tasks for his unit: evaluating which model would be best for initial operational use, preparing a program library, training personnel to program the computer, and training the analysts.

The Joint Numerical Weather Prediction Unit worked closely with the Air Force's Geophysical Research Directorate (especially Thompson's Numerical Prediction Project) and with the Institute for Advanced Study to evaluate models. Both the IAS and the GRD ran their three most promising models from the same initial data and compared the output. JNWPU members planned to obtain time on IBM's New York-based 701 to run some of the programs. By 1 February 1955, they anticipated running the three models with 30 different initial data sets. After studying about sixty of the GRD's baroclinic forecasts, the JNWPU's Development Section discovered that half of the systematic errors could be attributed to neglecting terrain-induced vertical motions. It was also analyzing the effects of ignoring some of the terms in the vorticity equation. Another study dealt with erroneous boundary assumptions and how they affected model output. However, they determined that 16 of the 60 500-mb-height forecasts were significantly more accurate than the subjective maps obtained from US Air Force Weather Central for the same verifying times. Based on these findings, the Development Section members revised the models and tested them by both hand and machine computation at the IAS. The Computing Section was working on a number of different programs, including barotropic, three-parameter baroclinic with terrain, objective analysis, and three-parameter baroclinic for comparison testing, as well as two-parameter baroclinic programs and a program that would produce a baroclinic forecast with boundary conditions given by a barotropic forecast covering a larger area.[84]

::

With the Joint Numerical Weather Prediction Unit officially open for business, the work of the Ad Hoc Group was at last done. However, the Joint Meteorological Committee wanted an oversight committee to work out disagreements among the three participating weather services. Therefore, on 4 November 1954, the JMC formally dissolved the Ad Hoc Group and established yet another ad hoc committee: The Ad Hoc Committee on Numerical Weather Prediction (referred to as JMC/NWP).[85] Each weather service was authorized to appoint one member, although others would be allowed to attend meetings in an advisory role. The JMC/NWP would stay cognizant of the workings of the JNWPU, and assist and advise its director on requirements, external technical matters, fiscal considerations, service personnel matters, and off-time equipment use. The JMC/NWP would keep the JMC informed of numerical weather prediction matters and bring any major policy questions to it for resolution. However, it was not within the committee's purview to solve any highly technical problems. For those, Cressman could seek the advice of scientific consultants after receiving the concurrence of committee members. This committee's quasi-supervisory role did not give it license to micro-manage the unit. As the director of a new unit in a new field, Cressman was to have significant freedom in determining what should be done.[86]

One of the first disputes, not surprisingly, concerned personnel. The Navy representative (Captain W. E. Oberholtzer Jr.) made clear that all Navy personnel assigned to the unit must be trained in each of its primary functions, i.e., modeling, programming, and analysis. Cressman responded that personnel would be cross-trained to the extent that there was a fit between their background and desires—not everyone wanted to perform all functions. Oberholtzer likely reacted badly, for Cressman's response pointed to a tremendous gulf between the cultures of civilian meteorologists and the military services: in the Navy, one's individual "desires" had nothing to do with one's assignments.

Cressman had already arranged for technical consultants. From outside government they included Charney, Gilchrist, and Bigelow from the IAS; Platzman from the University of Chicago; and Rossby from the University of Stockholm. (Rossby had been given an official role after 9 years of guiding the development of numerical weather prediction.) Consultants from within government service included computer specialists Lawrence Gates (of the GRD), von Neumann (now of the Atomic Energy Commission), and Franz Alt (of the National Bureau of Standards).

Another regulation was imposed: any agency desiring to place new requirements on the JNWPU had to coordinate them with JMC/NWP. Without this provision, chaos would have reigned immediately. The Air Force, the Navy, and the Weather Bureau, each having different mission requirements, were all seeking unique products. Without a clearing house for their specialized mission requirements, the unit would be overwhelmed. As far as requirements being levied *by* the JNWPU, Cressman reported that the WBAN Analysis Center would be plotting and analyzing two 400-mb charts per day starting in January 1955. Because the analysis section of the JNWPU had been kept small on purpose, he was counting on WBAN to fill its needs. Owing to the coordination required between JNWPU and WBAN, the ad hoc committee determined that the joint unit needed to deal directly with the Coordinating Committee of the National Weather Analysis Center (NWAC—replacing the WBAN Analysis Center) if and when the JMC appointed such a committee.

Also important was the policy for outside use of the IBM 701. The machine had not yet arrived, but outside agencies were already seeking computer time. The proposed policy stipulated that the computer could be used by governmental meteorological services or cooperating numerical weather prediction research groups subject to the director's approval. Computer use had to be at the unit's convenience, and it would provide no manpower other than a machine operator. Non-governmental groups using the machine had to pay for all computer time unless there was a reciprocal arrangement on another machine. When Wexler questioned why the JNWPU needed to be reimbursed when the machine time was already paid for, Cressman commented that they wanted to discourage non-meteorological organizations from using the computer.[87] The committee also discussed who would be allowed to submit programs to run on the machine. The Air Force representative thought first priority should go to whichever group had the most to contribute to numerical weather prediction regardless of whether it was a governmental agency or a research group. Regarding reciprocal computer time, Cressman noted that both the GRD and the IAS had run programs for the JNWPU, and therefore the joint unit should run programs for them if asked. They weather service representatives also addressed machine use in excess of contracted time. The cost increased once they exceeded the contract limit. Therefore, committee members decided that, as much as possible, any requests for computer time would have to fit within the contracted hours.[88]

With the beginning of fiscal year 1956 (1 July 1955) only 6 months away, the committee also considered its budget needs. Cressman anticipated no further staff increases after fiscal year 1956. He thought he might even be able to reduce staffing by one plotter. By the five-year point, he might reduce the programming staff. Apparently Cressman thought that, once they had programmed the models, little programming work would remain.[89] His casual comment was a consequence of a complete lack of experience in the field—no fault of Cressman, since everyone was new to the field—and yet it defied common sense. The unit's purpose was to take upgraded models and put them to operational use. The programming would always need to be done in-house. Therefore, the number of programmers would not decrease with time unless the models were not improved. The whole idea behind making the transition from a research to an operational organization was to speed up model improvements. Decreasing the numbers of programmers would risk stagnation.

The anticipated contribution of each service for fiscal year 1956 was approximately $200,000. The Air Force's representative advised that his service would need an estimated fiscal year 1957 budget not later than January 1955 (fiscal year 1957 would have started on 1 July 1956).[90] Somewhat surprisingly, military members did not push for budget estimates over several years, since their budgets were set up in five-year cycles and then readjusted each year as the operational climate changed.

In late 1954, Joint Meteorological Committee members addressed troubles coordinating with the new National Weather Analysis Center. Whereas the WBAN Analysis Center was supervised by JMC's Air Coordinating Committee/Meteorology (ACC/MET), the new analysis center would be without JMC supervision if a new ad hoc committee were not appointed to fill that role. On the other hand, the JNWPU fell under JMC's cognizance via the existing ad hoc committee. Therefore, the unit could not directly approach the analysis center for assistance; rather, it had to follow a cumbersome, circuitous chain through advisory committees *up* to JMC and *down* to the analysis center. This was clearly unsatisfactory. Since JMC had discussed merging the joint unit and the analysis center, members suggested that the JNWPU be placed under ACC/MET and the current ad hoc committee be dissolved.[91]

This matter resurfaced a few weeks later when JMC members discussed who should have supervisory authority over the joint unit and the analysis center. Although the Weather Bureau argued that there was no reason for the JNWPU to be under JMC supervision, all agreed that the best scenario would have both units being supervised by ACC/MET once they were operational. Since neither was operational, it was not of concern. In response to the Weather Bureau's comment about JMC supervision, the Air Force pointed out that it received part of its budgetary support for the joint unit because of its JMC affiliation. The Weather Bureau argued that unless it was absolutely necessary, no committee should supervise either unit because, as Reichelderfer put it, "committee operation of a unit is never good."[92] Undoubtedly Reichelderfer was partly motivated by the fact that both of these units resided in Weather Bureau spaces and were under its administrative control despite being jointly funded and staffed. The military units were probably concerned about losing control within a civilian organization without the JMC-related supervision.

The JMC discussed the use of computer time by outside agencies and concurred in the policy as proposed by the Ad Hoc Group. The JMC also brought up the fiscal year 1956 budget, but both military representatives asked for a deferral until they could study it. All agreed that they strongly supported the joint unit and did not anticipate failing to meet their share of the budget.[93] However, by mid January 1955, potential funding and manpower deficits were beginning to appear. The Air Force could not meet the manpower requirements, but could substitute funds for manpower even though it was not sure it could offer its full share. Although Weather Bureau leaders wholeheartedly supported the numerical weather prediction effort, the Commerce Department had eliminated the money they had set aside for the unit. However, the bureau continued to seek funding for its full share. The Navy reported that only part of its share had been included in its budget. That was partly due to the sharp increase in the unit's budget between fiscal years 1955 and 1956, as it moved from the pre-operational to the operational stage. The Navy needed to wait until the entire military budget had been adopted before knowing if there would be additional funds. The JMC then approved the proposed fiscal year 1956 budget with the stipulation that it would await the outcome of the total budgets of the Department of Defense and the Department of Commerce.[94]

Cressman briefed the JMC on the JNWPU's status on 3 May 1955, just 3 days before its dedication ceremony. The IBM 701 had been checked out and accepted from IBM 2 months earlier (figure 7.1). In mid April, unit members had run the first experimental forecasts with better than expected results.

Figure 7.1
Fred Shuman (left) and Otha Fuller at the IBM 701, circa 1955. (courtesy of NOAA-NCEP)

Since the computer had arrived later than expected, unit personnel would not complete the shakedown phase (Phase I) until 6 May 1955. At that time, unit members anticipated beginning Phase II operations. During Phase II, they would extend the objective analysis for North America approximately 1500 miles into the North Pacific and North Atlantic. This analysis would be for internal JNWPU use only. Unit members would also produce a baroclinic three-level prognostic chart for the United States. Difficulties with the introduction of terrain effects had led to some programming difficulties, but unit members expected to overcome those within a couple of weeks. They would also be producing vertical motion products between the 900-mb and 700-mb (3,300–9,000 feet) layers and the 700-mb and 500-mb (9,000–18,000 feet) layers at 12-hour intervals. The baroclinic and vertical velocity products would be available for users starting on 6 May. Work continued on a barotropic 500-mb prognostic chart covering all of the Northern Hemisphere.

The relative accuracy of the computer-generated charts generated a happy surprise for the joint unit. To verify the weather maps, unit members checked the 24-hour computer-produced prognoses at three levels (400, 700, and 900 mb) by comparing the distance between the forecasted and observed low height center positions. The 400- and 700-mb levels showed a difference of two degrees each of latitude and longitude, while the 900-mb level showed a 2° latitude difference and a 5° longitude difference. These results were better, Cressman argued, than the best subjective efforts, and should be considered as the worst possible outcome. These initial, experimental efforts were quite encouraging. Nevertheless, the handling of incoming data continued to be problematic. The data came in via teletype, and 15 man-hours later, unit members had finished manually punching the data onto cards and feeding them into the computer. The unit had obtained a machine that would read teletype paper tape and convert it to punched cards. This new procedure would reduce the number of sub-professionals from five to one. The computer could then be programmed to sort through the observations and reject reports that were either not needed for the objective analysis or were garbled before running the program. Cressman noted that automated data handling was just beginning, and it would be a number of months before this would become a routine.

The teletype system, sufficient for subjective methods, also was problematic. It took 9 hours just to collect all the data needed for a single chart. If the observation and transmission schedules were changed, it would take only 30 minutes. Adding up the time to tear teletype tape, punch cards, and run the programs, it was apparent that the 9 hours being absorbed by data collection would need to be dramatically reduced if numerical weather prediction were to work.

Military missions also affected data availability. For example, Air Force weather reconnaissance aircraft regularly flew and collected data at 500 mb (18,000 feet). The unit also created prognoses for that level, since it represented the steering level for surface systems and was highly valuable for forecasters. However, when the Air Force considered moving those flights to a higher level (400 mb) that meant the data stream would change as well as the desirability for producing 500-mb charts. If the Air Force did change its reconnaissance level, the unit might have to change its operations as well.

Cressman (pictured in figure 7.2) also wanted more data from over the Pacific. As a trial, the Weather Bureau had put an upper-air team aboard the USNS *General Hugh J. Gaffey* (a Military Sealift Command transport ship) while underway in the Pacific, and had obtained excellent results. The regular availability of soundings from ocean areas would help anchor the forecast. Another possibility: using dropsondes launched from aircraft transiting the area.[95] However, both of these were very expensive options, and if the bureau leaders were worried about having enough money to keep the unit operational, they probably did not have enough money to send upper-air teams out to ride ships of opportunity, i.e., ships transiting the area that were willing and able to take on men and material, or to send dropsondes out with military planes flying across the ocean.

Figure 7.2
George Cressman (left) was the first director of the Joint Numerical Weather Prediction Unit. Here, with Air Force Lieutenant Colonel Art Bedient, also of the joint unit, he prepares to mount a data tape for computer-microwave linking (circa 1960). (courtesy of NOAA-NCEP)

At Long Last, Dedication

By spring 1955 the Phase I shakedown was complete. Unit members had checked out the computer, personnel were on board, the model was running, and communications circuits were in place. The time had come for numerical weather prediction to leap beyond experiment and into operation.

The small, almost anti-climatic dedication ceremony took place at 5 p.m. on 6 May 1955—almost 9 years to the day since the Institute for Advanced Study had sent its proposal for a Meteorology Project to the Office of Naval Research. After a few short remarks and the ceremonial pushing of the "on" switch, attendees toured the facility for half an hour and returned to the computer in time to see the 12-hour forecast emerge from the printer. Thirty minutes later, the 24-hour forecast appeared. Operational numerical weather prediction, a dream since the early twentieth century, was a reality (figure 7.3).[96]

In his comments for the official opening of the Joint Numerical Weather Prediction Unit, Weather Bureau Assistant Chief for Operations Delbert Little paid tribute to the pioneers of hardware development, upper-air investigations, dynamic meteorology, and, of course, to Lewis Fry Richardson, who in 1922 published the less than promising results of his attempt at numerical

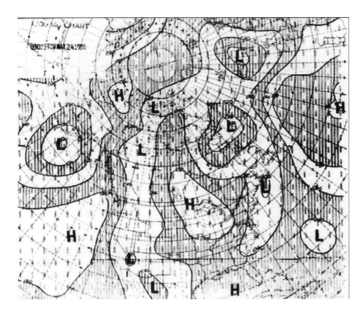

Figure 7.3
A 1,000-millibar prognostic chart, 24 May 1956, created by the JNWPU's IBM 701 using a grid-point contouring program written by Art Bedient. (courtesy of NOAA-NCEP)

weather prediction. And credit was given to von Neumann and Charney for their leadership in the Computer and Meteorology Projects and for bringing to fruition two of the three legs on which numerical weather prediction stood: the electronic computer and meteorological theory of large-scale atmospheric motions. The third leg—a sufficiently dense network of surface and upper-air observations—was in place as a result of World War II. Credit was also given to Air Force, Navy, and Weather Bureau personnel who had played important roles in the development and planning of this "unprecedented venture." Absent was any mention of the tag team of European meteorologists—primarily Scandinavians and Britons—who had bailed the Meteorology Project out of numerous manpower holes. These meteorologists, who had stayed in the United States for several months to a year at a time, had been crucial to creating the dynamic-synoptic meteorology interface required for the successful creation of numerical weather prediction models. They have been omitted from accounts of numerical weather prediction's origins ever since.

In his speech to the handful of Air Force, Navy, and Weather Bureau officials and a few invited guests, Little emphasized that the "new era in meteorology" that provided these computer products was not an excuse to "sit back and take it easy." On the contrary, forecasters would now have more time to devote to their local forecasts, with the computer taking care of the large-scale forecast. Modeling results had revealed that topographic, coastal, and diurnal effects were more subtle than previously thought. This discovery would allow meteorologists to concentrate their efforts on other elements that might ultimately prove to be more important to solving the forecasting problem. The computer "under intelligent human direction" would be the forecaster's assistant—not the controlling factor in making forecasts.[97]

And so it was. Or was it? How acceptable would these computer products be to the men at the forecasting desk? How would they be viewed by the different weather services? What was the future of the resulting man-machine interface? How long would it take before this operational unit was truly operational? How long could a melding of personnel from different operational backgrounds, serving very different customer bases, function before infighting led to its disintegration?

8 :: A New Atmosphere

If the Meteorology Project was the cradle of numerical weather prediction—the initial research phase required to put numerical methods on a firm theoretical footing—then the Joint Numerical Weather Prediction Unit might be considered the "baby walker." It took the project's models into the operational phase, putting them to the test daily, with real-time data used to make real-time forecasts. But while the theory was relatively firm, the operations were decidedly shaky.

The Joint Numerical Weather Prediction Unit did not survive as long as the Meteorology Project that spawned it. Long-term cultural differences and unique customer requirements contributed to its demise, but the inclusion of numerical weather prediction techniques on the forecasting floor lived on. Not wholeheartedly embraced at first, numerical products became the foundation of operational forecasting, including that done for television newscasts, by the 1970s.

Although the Meteorology Project faded away—its work did not fit into the Institute for Advanced Study's vision once John von Neumann decamped to the Atomic Energy Commission—theoretical numerical weather prediction spread to newly established research groups and university campuses around the world. Meteorologists, formerly concentrated within meteorology organizations and academic departments wealthy enough to lease mainframe computers, now formed alliances with other disciplines, incorporating new techniques and data sources into increasingly complex models. By the end of the twentieth century, modeling had expanded temporally and spatially to include climate and "earth-system" forecasting. As concerns about weather and climate changes increased, so did numerical efforts. The return to collaborative efforts in the twenty-first century—driven once again by the need to share limited resources—brings 50 years of numerical weather prediction full circle, and will likely lead to interesting tensions among those providing funding, equipment, and personnel to solve problems involving the earth and its atmosphere.

Culture and Customers Trump Jointness

In 1955, the JNWPU was an operational entity in only the loosest sense. Computer-generated products did not leave the unit, decisions about appropriate models remained to be made, and model development continued inside and outside the unit. Moreover, lots of rough edges still needed smoothing, including hardware use, data handling and coding, and transmission schedules. With these elements unresolved, it was realistic to expect that a joint organization combining the talents of Air Force, Navy, and Weather Bureau meteorologists, mathematicians, programmers, and subprofessional assistants would lead to the quickest results. Indeed, until meteorologists were forced to produce numerical forecasts on a schedule—i.e., before the forecast period started and not weeks or months later—they would have little impetus to improve model efficiency or solve long-term difficulties related to obtaining and handling meteorological data. Once some of these basic matters were under control, however, the situation began changing for the three collaborating weather services; the beginning of the end for joint operational numerical weather prediction was evident within a year of the unit's opening.[1]

Despite attempts by various government watchdogs to root out and eliminate overlapping weather services during the first half of the twentieth century (and into the second half), an indisputable fact was that the Army Air Force (later the US Air Force), the Navy, and the Weather Bureau forecasted weather for very different customers. Furthermore, these organizations possessed very different cultures that influenced their practices, and the cultural differences had doomed previous efforts to force them to work together.

Disagreements foreshadowing the JNWPU's 1960 demise appeared within 6 months of the unit's opening. There were four basic points of contention: model selection, model coverage (geographic and spatial), replacement of hand-drawn products, and transmission of computer-generated products on the facsimile broadcast. Each service examined these points and developed a unique approach.

The Weather Bureau, which exerted administrative control over the unit, focused primarily on its ground-based customers in the continental United States. It supported aviation interests, but not high-flying, high-performance aircraft; the Air Force forecasted for them. The Weather Bureau supported marine interests, but not ships at sea; the Navy issued forecasts for them. Moreover, the Weather Bureau's leaders were not about to provide the United States with computer-generated products unless they were *superior* to extant subjectively produced weather forecasts. *By law* they were responsible for providing the nation's weather services, and therefore would provide the best possible analyses and prognoses by the best available method—computer or hand-drawn.[2] This pragmatic, civilian-dominated mindset influenced every Weather Bureau decision. In contrast, the Navy's customers were at sea or in the air nearby. To serve them, Navy meteorologists needed weather maps that covered the vast extent of the Atlantic. Therefore, they needed to assess and include ocean-based data sources in the models. They also needed to expand the models' geographic extent. Until oceanic coverage improved, the Navy argued, the JNWPU's prognoses were of "academic interest only" and "of little or no operational value." Navy ships receiving US-centered charts via facsimile were not getting operationally relevant information. The JNWPU could not accommodate the Navy's needs without increasing its manpower authorization by 30 percent, and it was already 25 percent undermanned. The Navy was not going to get its desired products.[3]

The Air Force provided forecasts for ground forces, but its primary customer was airborne and he was *way* up there. Therefore, it needed numerical products that provided forecasts for high-altitude flight. Air Weather Service meteorologists wanted to see a multi-level baroclinic model in place, the sooner the better. More critically, the Air Force wanted to see numerical research results incorporated into JNWPU models as soon as they were available. By December 1955, Air Force meteorologists were demanding closer ties with the Air Force Geophysical Research Directorate.[4]

But while the Weather Bureau accepted multi-level baroclinic models, the Navy did not. Barotropic models worked best over the Eastern Atlantic, where Navy ships operated. Shifting to an operational baroclinic model threatened to leave Navy meteorologists in a bind. They were already experiencing difficulty forecasting in the Pacific owing to lack of data. If they could not use the Atlantic charts, then numerical weather prediction was doing nothing for them. Getting testy in the spring of 1956, Navy leaders wanted outside reviewers to ensure model selections were made "without bias." The data to support these models—very-high-altitude reports for the Air Force and additional oceanic reports for the Navy—were being debated for the same reason. Money was tight. Who would get the data important to them?[5]

As the intensity of inter-service skirmishes waxed and waned, technological deficiencies also plagued the JNWPU. The IBM 701, installed with great fanfare just the year before, was quickly overwhelmed by models that almost exceeded its capacity. The JNWPU could not get a new IBM 704 for at least a year, and unit members were aware that their models would use the machine's entire capacity shortly after its installation. Furthermore, computer costs would double in a time of fiscal conservatism. If the weather services did not stick together, the Weather Bureau was not sure they could all move forward. George Cressman argued that efforts to expand meteorological computing should be made "as part of one United States *system* and such expansion should receive careful *joint* study and action."[6] Cressman voiced his concerns in June 1956, well before any of the weather services openly discussed pulling resources out of the JNWPU.

The JNWPU's disintegration in the late 1950s can be based partly on different styles and approaches to modeling. In particular, the Weather Bureau's and the Air Force's theoretical approach clashed with the Navy's ad hoc, "can do" approach. A model's theoretical correctness was of little concern to the Navy, which just wanted acceptable products for its operational forces. Yet the most widely publicized reason for the split involved the content of and control over the facsimile broadcast. Weather maps were sent on a schedule—the same chart type was transmitted at the same time each day. The broadcast schedule was usually full, so adding a map meant pulling off an existing map. Since the Weather Bureau controlled the broadcast, its leaders viewed any attempt to add a new weather map, substitute a computer-produced map for a hand-drawn map, or pull a weather map from the schedule as meddling in a bureau responsibility. Navy and Air Force leaders did not think requesting schedule changes usurped the Weather Bureau's authority. If, after consultation, they all agreed with a Weather Bureau decision on the broadcast schedule, everyone was satisfied. If they did not agree, the military weather services wanted the final decision to rest with the Joint Meteorological Group (formerly Committee). After all, they argued, the Navy and the Air Force were providing a considerable share of the funding, and these maps should support their unique missions. For its part, the Weather Bureau thought Air Force personnel were "unfamiliar," "not well briefed," and "suspicious and overly cautious" about numerical weather prediction. Perhaps, but the Navy agreed with the Air Force. So did the Joint Meteorological Group. George Cressman and other Weather Bureau meteorologists were unhappy. In their eyes, the military weather services were moving into operational and research areas outside their purview. The Weather Bureau had recognized that within the Cold War context more funding would go to military than to civilian meteorology. The Weather Bureau had gamely marched on, but its patience had worn thin. Francis Reichelderfer thought the responsibilities of each service should be clearly delineated so funding could be adjusted accordingly.[7]

When the budget for fiscal year 1961 came up for review, the Navy suggested that the Weather Bureau fund the entire operation, which had amounted to $1.7 million in 1960. The Air Force agreed, while indicating that it would not pull out immediately.[8] The Navy argued that military budgets were uncertain, and it did not want to make a commitment. Military budgets were indeed uncertain, but rarely so uncertain that the cost of providing weather forecasting services could not be covered. The Navy had been happy to participate in cost sharing when the JNWPU was basically a research organization. Now that it was operational, the Navy argued that the Weather Bureau should fund numerical weather prediction. Although this round of budget disputes was overruled by the Bureau of the Budget, which forbade the parties to remove funding, the JNWPU was close to disbanding.[9]

The end came in the waning days of the Eisenhower Administration. In January 1961, the Navy announced that it was opening a Fleet Numerical Weather Facility on the grounds of the Naval Postgraduate School in Monterey, California. The facility's stated mission was to provide operational numerical weather prediction products peculiar to the US Navy's needs, including the development and testing of numerical techniques in both meteorology and oceanography.[10]

The Air Force had already established its own computer facility at Offutt Air Force Base, Nebraska, and was processing classified weather products separately. By late summer 1961, what remained of the JNWPU was subsumed under the Weather Bureau's National Meteorological Center (NMC). After 15 years of cooperation on numerical weather prediction, the three weather services struck out on their own paths.[11] Numerical products would henceforth be created by models unique to their respective organizations. But the attitude on these three services' forecasting floors shared a common characteristic: resistance.

Numerical Weather Prediction on the Forecasting Floor

In the early days of numerical weather prediction, theoreticians and meteorologists who became modelers and/or programmers were the only people affected by the new way of doing meteorology. But once computer-created predictions hit the forecasting centers, everyone—including observers, plotters, and forecasters—had to adapt. It is safe to say that most operational forecasters in the late 1950s and the early 1960s described the early computer-generated predictions (figure 8.1) in scatological terms. And they were right. Early computer charts were more often than not wadded up and thrown away as forecasters turned back to the hand-drawn products with which they felt comfortable. It is quite possible that forecasters would never have incorporated computer-generated maps into their routines except for one very important decision: the hand-drawn charts were withdrawn from the facsimile broadcast and replaced with numerical products. The only

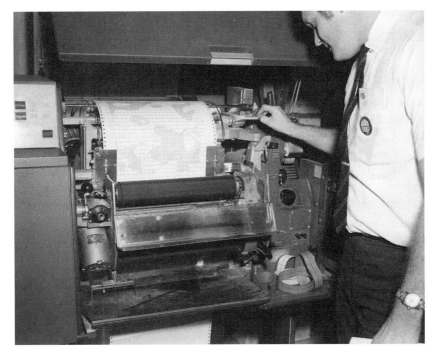

Figure 8.1
A computer-produced weather map appears on a line printer at the National Meteorological Center, circa 1962. (courtesy of NOAA Central Library)

choice was to mutter, grumble, use them, and provide the feedback modelers needed to identify deficiencies and correct them.[12]

Operational meteorologists had concerns in addition to map quality. Once automatic data-handling processes became more sophisticated, the charts showed up pre-plotted with observations. That meant that plotters—who had to be able to quickly record (in ink) an entire station plot (coded observation data) within a dime-sized area—had very little to do. Military service members still had to plot classified reports transmitted over encrypted lines, but there was a huge difference between plotting a few classified ship or aircraft observations and plotting observations over an entire continent. But since computers and their associated systems were prone to failure, plotters still needed to retain their speed and accuracy so that forecasters would have enough to time to analyze the current surface map and get forecasts out on time. Plotting remained a requirement in most operational settings through the 1980s and into the 1990s, as did hand-drawn surface and upper-air analyses—both as a way to train new people and as a way to provide model feedback to numerical weather prediction centers.

When forecasts had been based on longstanding rules of thumb and on years of experience watching similar weather patterns play out, operational forecasters had not been forced to retool their techniques. That changed with the arrival of numerical weather prediction and frequently updated models. Every time programmers updated a model—events announced as being "transparent to the user" that were anything but—forecasters had to adjust their understanding of the model's strengths and weaknesses. The ability to make that adjustment depended on the forecasters' understanding of the atmospheric situation, which in turn depended on their experience. Sometimes even an uninitiated person could have determined at a glance that a given model run had failed—a giant "bull's eye" in the middle of the forecast map yielded plenty of undeleted expletives about numerical products on the forecasting floor. But once the members of a forecasting team became accustomed to a model's latest quirks, they knew when they could trust its output and when they could not. Forecasters' knowledge of model biases, typically encountered with particular weather regimes, provided the necessary human "sanity check" on numerical predictions. Operational forecasters also provided modelers with the information they needed to increase the accuracy and the temporal range of weather predictions in a feedback loop that tied theoretical modeling and operational expertise together.

As slow as the transition was from all subjective to all objective analyses and prognoses on the forecasting floor during the 1960s and the 1970s, it was quick compared to the glacial rate of change in how forecasting teams handled computer-generated weather maps. No matter how weather maps were created—by hand, main-frame computer, or desktop workstation—they invariably ended up posted on a wall so the forecast team could assimilate the current and future atmospheric states. Although meteorologists entering the field since the introduction of workstations (for example, the National Weather Service's AFOS; see figure 8.2) in the 1980s seem to deal better with examining maps on screens, many field forecasters still find it easier to absorb the output from the rapidly expanding suite of models if maps are printed out. New generations of super-computers provide the opportunity to produce longer-range forecasts for atmospheric layers and variables. Operational forecasters using these products continually adjust the forecasting strategies they use to stay ahead of the ever-changing atmosphere.

Figure 8.2
Richard Hallgren of the National Weather Service formally accepts delivery of the Automation of Field Operations and Services (AFOS) system while sitting at an AFOS console in February 1978. (courtesy of NOAA Central Library)

Operation Push and Theory Pull

In the decades following the Meteorology Project and the JNWPU, model development and operational products burgeoned. The Air Weather Service, the Naval Weather Service, and the Weather Bureau (and their successors) all built increasingly powerful computing centers and developed distinct models to serve their customers. Operational modeling centers expanded overseas too—first in efforts confined to individual countries and then to large collaborative efforts such as the European Center for Mid-Range Weather Forecasting (near Reading, England), which has grown from 16 to 30 member nations since 1975. With each new generation of computers, models included more variables and different approximations, time steps, and geographic areas. Grid meshes became finer and time periods extended from a few days for weather forecasts to centuries for climate predictions. A combination of theoretical and operational work contributed to the improvement of models.[13]

As the Meteorology Project and its theoretical modeling ended, the wider academic community took up the development of atmospheric models at the Weather Bureau's Geophysical Fluid Dynamics Laboratory (later associated with Princeton University), the National Center for Atmospheric Research in Boulder, Colorado, and in university atmospheric sciences departments and institutes all over the world. How these groups came into existence is an important sequel to the efforts to develop numerical weather prediction.

John von Neumann and Jule Charney had encouraged Francis Reichelderfer to establish a group separate from the JNWPU that would focus its research on the general circulation of the atmosphere—work that had been started by Norman Phillips of the Meteorology Project.[14] Obtaining additional funding from the Navy and the Air Force, Reichelderfer and the Weather Bureau established the General Circulation Research Section in 1955 to conduct basic research on the general circulation of the atmosphere and on climate. Joseph Smagorinsky became its director in October 1955 and remained its director through two name changes until his retirement in 1983.[15]

The five-member staff began its work on general atmospheric circulation, but Smagorinsky soon realized that if they were to extend their efforts to climate modeling they would have to incorporate ocean modeling, which was first included in 1960. Until that time, modelers had assumed a static ocean. Indeed, early models omitted surface interactions because they were too difficult to model—another source of the Navy's discontent with model behavior over the oceans. Smagorinsky also realized the importance of modeling individual meteorological processes, as well as taking the lessons learned from those virtual experiments and applying them to weather forecasting and climate models. Pioneering interdisciplinary interactions in meteorological modeling, the research group gradually included radiation, condensation, boundary layer, and ocean processes in their increasingly complicated models. By 1963, the group had a new name (Geophysical Fluid Dynamics Laboratory) to go along with its expanded mission. Five years later, the GFDL moved to Princeton to become associated with Princeton University. The interdisciplinary nature of the GFDL's modeling efforts is illustrated by the departments with which it maintained both formal and informal ties: civil engineering, aerospace and mechanical sciences, geology and geophysical sciences departments, as well as astrophysics, chemical engineering, and statistics. In the mid 1970s, the GFDL became affiliated with Princeton's Geology and Geophysical Sciences Department.[16]

The Geophysical Fluid Dynamics Laboratory tested atmospheric circulation models by using real-time initial data, running the models as if they were forecasts, and checking the computer output against observed data. In 1963, Smagorinsky published his primitive equation model of the general circulation—a significant breakthrough in model sophistication.[17] The primitive equation model had been rejected in the 1940s and the 1950s in favor of simpler models, but advances in computing power in the 1960s made modeling the primitive equations possible for the first time. The GFDL team later increased model run times and turned its attention to modeling wind-driven and heat-driven ocean circulations, convection, hurricanes, mesoscale features, and climate.[18] The weather and climate models developed by the GFDL have significantly increased meteorologists' understanding of atmospheric circulation, including heat and moisture transport, and the impact of humans on earth's changing climate. As operational computing centers upgraded to more powerful computers, they incorporated the results of the more theoretical investigations conducted by the GFDL and other research centers.

Not only did the widespread acceptance of numerical weather prediction change the approach to theoretical meteorology and operational predictions; its need for massive amounts of data inspired large international efforts to coordinate data gathering. Jule Charney of MIT considered the obstacles to obtaining meteorological data and started mulling over ways to increase international cooperation in atmospheric science. He proposed a major international data-gathering experiment. Later known as the Global Atmospheric Research Program (GARP), Charney's scheme for pulling together massive amounts of observational data from every conceivable direct and remote sensing platform got started in the late 1960s.[19] Its main event—the $500 million year-long Global Weather Experiment of 1978—involved scientists from 147 nations gathering data from satellites, buoys, ships, and constant-level balloons.[20]

An alphabet soup of data-gathering exercises in data-deficient areas followed. The availability of increasingly sophisticated satellite-based sensors has allowed atmospheric scientists to collect more detailed information on areas that are outside the reach of standard surface and upper-air stations. Automatic observing stations on land and afloat also provide a steady stream of data to some of the world's largest computing facilities, enabling scientists to conduct interdisciplinary "earth system" studies such as were not even dreamed of by numerical weather prediction pioneers in the 1950s.

Numerical "Earth System" Prediction

As sensing systems multiply, assimilation systems collect and process even more data, and sophisticated models are created to handle them, numerical prediction will increasingly incorporate more "earth system" modeling than atmospheric modeling. Forging an interdisciplinary approach, modelers will link climate, weather, water, land, cryosphere (ice), space weather, and chemistry to expand prediction systems to include environmental forecasts of air and water processes, and ecological processes. Weather predictions will be enhanced by combining ocean, land, and atmospheric forecasts. The effects of solar activity on weather and climate—dismissed in the early twentieth century—are once again being incorporated in models. By combining air chemistry and operational models, ozone forecasts will be improved. Operational forecasts—once limited to temperature and precipitation—will expand to include ocean conditions, ecosystems, the carbon cycle, and solar flares. Numerical models are not likely to provide solutions to every atmospheric and broader "earth system" problem, but they will continue to be exploited as a practical way to experiment on earth's processes that cannot be confined to a laboratory bench.[21]

How will this work be coordinated? Collaboration such as existed in the 1940s and the 1950s among the Air Weather Service, the Naval Weather Service, and the Weather Bureau is being resurrected by their successor representatives (the Air Force Weather Agency, the Fleet Numerical Meteorology and Oceanography Center, and the National Centers for Environmental Prediction), which have signed a national Weather Research and Forecast framework. The Weather Research and Forecast framework—an initiative of the US Weather Research Program that was formed to foster partnerships among federal agencies, universities, and a variety of research institutions—is a collaborative effort among operational prediction agencies and the research community whose goal is to put research results into immediate operational use. Like the Joint Numerical Weather Prediction Unit, it provides a structure for civilian and military operational weather agencies to share resources, increase efficiency, and provide "earth system" support for their users. The participants hope the newly developed numerical products will allow operational forecasters time to provide expert scientific advice to decision makers and to the public, particularly in the days leading up to and during hazardous weather events. In view of the dramatic changes in operational meteorology in the last 50 years, it is fairly safe to predict that another 50 years of theory, model, and computer development will produce even greater changes in how "earth system" scientists predict and explain changes in the earth's environment.[22]

Conclusion

The introduction and acceptance of numerical weather prediction was the great watershed of twentieth-century meteorology. Numerical weather prediction provided a technique and a way of thinking that represented a fundamental discontinuity in meteorological theory and practice and transformed meteorology from a marginal scientific backwater to a rigorous science at the forefront of scientific discourse within 20 years. Particularly in the United States, where meteorologists were not often considered "real scientists" by their colleagues in the physical sciences, numerical weather prediction became the mark of a professional scientific discipline. Due to numerical weather prediction, meteorology developed a robust theory, established a strong academic presence, experienced a rapid increase in the availability of research funds, and saw graduate-school-trained meteorologists remake the nation's weather services into respected, professional organizations.

As important as the story of meteorology's twentieth-century professionalization is, the Meteorology Project is an appropriate subject for the study of a number of themes in twentieth-century science: military funding and civilian applications, the integration of practice with theory, the tensions between military and civilian control of science, the question of what constitutes acceptable scientific evidence, and how to predict natural phenomena (past, present, and future) using ever more sophisticated computer models.

The Meteorology Project serves as a unique example of a military-funded effort leading to significant scientific advances that had immediate civilian applications. During this same period, the Office of Naval Research was also funding extensive oceanographic research, which obviously aided the personal research agendas of civilian oceanographers, but did not have a civilian payoff equivalent to weather forecasting.[23] Nor did the military-funded space sciences, which captured the public's imagination but did not affect daily life.[24] Although a large project by meteorology standards, the Meteorology Project was very small compared to the Manhattan Project and the large particle accelerator facilities that characterized physics by the middle of the twentieth century. And yet the "small science" Meteorology Project led to the meteorology's transition to "big science" within a decade. And unlike many of the Cold War military-funded physics projects, it did not face compartmentalization and secrecy—indeed, information about the project's mission and what it would do for the average citizen were spread by the media. The Meteorology Project also provides insights into the integration of the electronic computer into scientific research, one of the most significant developments in the history of post-World War II science.

The Meteorology Project is also an important example of how practice was integrated with theory in mid-twentieth-century science. Although theoretical meteorologists (dynamicists) took the lead in starting the modeling effort, meteorologists who combined theory and practice (synopticians—especially the Scandinavians) were crucial to its ultimate success. Before their arrival, models were, as Jule Charney lamented, veering toward "mathematically sterile" entities that were internally consistent but had little relation to atmospheric reality. The models could produce descriptions of atmospheric layers that were correctly configured in relation to all the other modeled layers, but failed to match the atmosphere. Why this lack of connection between model and reality? In the United States there was an almost insurmountable intellectual, professional, and institutional gap between those who developed meteorological theory (academic dynamicists) and those who practiced meteorology (that is, the weather services' forecasters)—a "valley of death" that still existed at the start of the twenty-first century. Without a common language, American researchers and operational forecasters were unable to collaborate on a project that was mutually advantageous. Recall John von Neumann's words as he attempted to attract a Scandinavian synoptician to the project: "[This] kind of cooperation . . . will ultimately have to take place between theoretical and synoptic meteorologists, if and when numerical forecasting is integrated into the governmental weather services."[25] More than 50 years later, the statement remains true.

The Meteorology Project is also illustrative of tensions between military and civilian control of science—another important theme in twentieth-century American science. During World War II, meteorological services had quickly swung to military control. The prewar relationships between the military weather services and the Weather Bureau were not immediately resumed in the postwar years, as institutional missions remained in flux. However, it was the Weather Bureau's Reichelderfer who recognized the potential of computer-assisted weather prediction, along with his organization's inability to fund it. The money would have to come from the military services, which had come to dominate both operational and research meteorology in the name of national defense. Reichelderfer (a retired naval officer) understood this military necessity, but as the Meteorology Project moved forward he was determined to reclaim the Weather Bureau's traditional role as the nation's provider of weather services and to ensure that the bureau and its customers were not excluded from the meteorological research agenda.

Supervision of the Weather Bureau had shifted from the Agriculture Department to the Commerce Department in 1940 as aviation interests had replaced agricultural interests as the customers with the loudest voices.[26] Indeed, aviation mission requirements were the reason the military basically co-opted meteorology during World War II. And then as now, the organization that controlled the funding also controls the research agenda. Military meteorology had no interest in broader agricultural applications of meteorology—a strictly civilian function—and in time would spend more of its money on military applications while the National Science Foundation funded basic theoretical work. To regain control of at least part of the meteorology research budget, the Weather Bureau would have to make a strong argument for the economic importance of advancing agricultural meteorology. The reasons behind military and civilian meteorology research funding decisions from the 1950s on need closer historical examination.

The story of the Meteorology Project also illuminates a significant epistemological point that seems characteristic of the twentieth century. For meteorology, it can be stated as follows: Who possesses the better understanding of the atmosphere—those who crunch numbers but never look outside, or those who are unimpressed by equations but read the sky? This is perhaps analogous to the dispute between geophysicists and field geologists—the former as the number crunchers and the latter as those who bring back "solid" evidence—as to who had the best understanding of geological reality.[27] It is also related to the professionalization of scientific communities. Who is considered to be the "real" professional who speaks for the community? Who is the technician—allowed to work within the community, but never to speak for it? This last point was, in part, what triggered the in-fighting that overshadowed the transition from research to operations in numerical weather prediction. On some level, the warring participants forgot that the Meteorology Project had been successful by virtue of being "joint." The project could not have reached its ultimate goal of creating a theoretical basis for meteorology without the participation of every meteorological constituency. The military services and the Weather Bureau—the project's fiscal and technical supporters—likewise could not have reached their goal of faster, more accurate weather forecasts for their customers without the participation of their academic brethren.

Last, the Meteorology Project was the first instance of a major development that came to shape late-twentieth-century science: the use of computers to model nature and the behavior of natural systems. By 2000, computer models had become commonplace yet powerful tools for forecasting processes across the natural sciences, including past and future changes in ecological systems and earth's climate. Sophisticated computer programs also allowed weapons specialists to anticipate the physical behavior of aging nuclear weapons without the need to test them through actual detonation.[28] Some of these applications were controversial when first introduced, and some remain controversial; what is important to remember here, however, is that numerical forecasting in meteorology illuminates how computer-based modeling was first used to assess and understand complex natural systems.

The story of this achievement is part of the history of the rise of modeling in the natural sciences.

Those involved in the development of numerical weather prediction certainly valued the power of the tool they were creating and the challenges of making it part of the normal practice of science. During the nearly 9 years of research leading to operations, the project meteorologists came to realize that they would not win over their theoretical colleagues to numerical weather prediction's potential power if they could not show a connection between their equations and the atmosphere. To convince the skeptics, they would have to use their models to predict the weather. The applied meteorologists, searching for predictive assistance, knew that they would never achieve their goal unless they understood which atmospheric variables were truly critical to the forecasting process. Overcoming backbiting, bruised outsized egos, unbridled ambition, longstanding inter-service rivalries, manpower shortages, and at times almost overwhelming discouragement, they ultimately triumphed with the major advance in twentieth-century meteorology: the development and acceptance of numerical weather prediction.

Notes

Archival Sources

- Library of Congress, Manuscript Division, Washington, DC
Edward L. Bowles Papers (abbreviated as Bowles)
Francis Wilton Reichelderfer Papers (abbreviated as Reichelderfer)
Harry Wexler Papers (abbreviated as Wexler)
John von Neumann Papers (abbreviated as von Neumann)

- Institute Archives and Special Collections, MIT Libraries, Cambridge, Massachusetts
Jule Gregory Charney Papers, MC 184 (abbreviated as Charney)
University Meteorological Committee Papers, MC 511 (abbreviated as UMC)

- National Academy of Sciences Archives, Washington, DC
Executive Board, Science Advisory Board, Committee on Weather Bureau (abbreviated as SAB) NAS Organizations 1938–1939 (abbreviated as NAS/SWB)
Agencies and Departments

- California Institute of Technology Archives, Pasadena, California
Robert A. Millikan Papers (abbreviated as Millikan)

- National Center for Atmospheric Research Archives, Boulder, Colorado
Philip D. Thompson Papers (abbreviated as Thompson)

- National Archives and Records Administration II, College Park, MD
Weather Bureau Papers, RG 27 (abbreviated as USWB)

- Manuscripts, Special Collections, and University Archives, University of Washington, Seattle
WU President, Accession 71-34 (abbreviated as WU President)

Notes to Introduction

1. Francis W. Reichelderfer to Weather Bureau Staff Members, 29 December 1945 (Wexler, 2/45).

2. Quoted in Koelsch, "From Geo- to Physical Science," 526.

3. Richardson, *Weather Prediction*, 219. Citations are of the Dover edition.

4. "Curiosities of Science and Invention: Meteorology in American Universities," 343.

5. Bates and Fuller, *Weather Warriors*, focuses on the military weather services and weather's effect on military operations, but fails to address military contributions in advancing the discipline. See also Fuller, *Thor's Legions,* for an operational history of the US Air Force's Air Weather Service.

6. Charles F. Sarle memo, 15 January 1942 (Wexler, 1/Gen. Corr. 1942).

7. See Nebeker, *Calculating the Weather* and Aspray, *John von Neumann.*

8. See Geison and Holmes, *Research Schools.*

9. Justification Memorandum, PD #EN1-22/00028, The Institute for Advanced Study, 6 June 1946 (von Neumann, 15/6).

10. For a detailed look at Bjerknes and the Bergen School, see Friedman, *Appropriating the Weather*.

11. Brooks, "Reclassification," 163–164; Cox, "Curious Ways," 54–55, 150.

12. Palmer, "Weather Insurance," 67–70.

13. Unpublished table of the National Research Council, Table III Doctorates Conferred According to Subjects (1923) (NAS, Research Information Service, 1920–1923, Information Files Doctorates Conferred).

14. Ward, "Meteorological Instruction," 554.

15. Brooks, "Our Society's First Decade," 8–12. See Kevles, *Physicists*, and DeVorkin, *American Astronomical Society's First Century*, for discussions of physics and astronomy professional societies.

16. C.-G. Rossby to Major General B. M. Giles, 10 September 1943 (Bowles, 30/4).

17. Byers, "Recollections," 217.

18. Harry Wexler to Francis W. Reichelderfer, 4 November 1946 (Wexler, 2/1946).

19. Jule Charney to Rossby, ca. 19 March 1947 (Charney, 16/516).

20. On national styles in science, see Nye, "National Styles?"

21. Charney to John von Neumann, 2 January 1948 (von Neumann, 15/1).

22. The Institute for Advanced Study, The Meteorology Project, Summary of Work, under contract N-6-ori-139 (10), NR 082-008 during the calendar year 1952 (Charney, 9/304).

23. Norman Phillips to Charney, 15 September 1952 (Charney, 14/449).

24. Reichelderfer to Project Leader (Joseph Smagorinsky), 19 February 1954 (Wexler, 6/1954).

25. Phillips's work on the general circulation models was a precursor to climate modeling; see Phillips, "General Circulation." The development of climate modeling in the twentieth century is beyond the scope of this book. On the emergence of climate models, see Edwards, "Representing the Global Atmosphere." Human recognition of possible anthropogenic contributions to climate change is another distinct issue; see, for instance, Weart, "Nuclear Frying Pan"; Fleming, *Historical Perspectives*; and Weart, *Global Warming*.

Notes to Chapter 1

1. Kimball, "Recent Advances," 4.

2. Van Hise, "Report," 1. See also Good, *The Earth, the Heavens*.

3. See, e.g., Goodstein, *Millikan's School*, and Geschwind, *California Earthquakes*.

4. On philanthropy and the natural sciences, see Kohler, *Partners in Science*; on the creation of the Scripps Institution, see Rainger, "Adaptation and the importance of local culture"; and on geophysics at Harvard see Walter, *Science and Cultural Crisis*.

5. See, e.g., Doel, "Geophysics in Universities"; for a review of geophysics training in the United States, see Straley et al., "Professional Training," 398.

6. Jones, "Annual Report" of the Director, US Coast and Geodetic Survey, to the Secretary of Commerce for the Fiscal Year ended June 30, 1922," quoted on 7.

7. Kimball, "Recent Advances," 4, 6.

8. Gregg, "History of the Application of Meteorology," 165.

9. Brooks, "Reclassification," 164.

10. 26 *Statutes at Large* 653, section 3 (1890). See Whitnah, *History of the United States Weather Bureau*, 131–200, for a discussion of this period in Weather Bureau history, based primarily on its internal publications and the transcripts of legislative hearings. Fleming, *Meteorology in America, 1800–1870,* is the definitive work on these early years of meteorology in the United States. Kutzbach, *A Thermal Theory of Cyclones,* addresses the development of meteorological thought in the United States and Europe from the mid-nineteenth century to the polar-front theory of the Bergen School.

11. Weber, *Weather Bureau*, 17, 44.

12. US Weather Bureau, *Report of the Chief, 1932–1933*, 1.

13. Weber, *Weather Bureau*, 17.

14. US Weather Bureau, *Report of the Chief, 1920–1921*, 19; US Weather Bureau, *Report of the Chief, 1921–1922*, 14–15, 17; US Weather Bureau, *Report of the Chief, 1923–1924,*

8; US Weather Bureau, *Report of the Chief, 1925–1926*, 4. See Pyne, *Fire in America*, 314–317, for background on fire issues and a discussion of fire weather forecasting as provided by the Weather Bureau (later the National Weather Service) and the coordination that took place with the US Forest Service.

15. Pyne, *Fire in America*, 317.

16. Cox, "Curious Ways," 54.

17. Weber, *Weather Bureau*, 70.

18. US Weather Bureau, *Report of the Chief, 1920–1921*, 11.

19. Ibid., 13; US Weather Bureau, *Report of the Chief, 1933–1934*, 3; US Weather Bureau, *Report of the Chief, 1921–1922*, 27; US Weather Bureau, *Report of the Chief, 1922–1923*, 10.

20. US Weather Bureau, *Report of the Chief, 1920–1921*, 10, 16; US Weather Bureau, *Report of the Chief, 1921–1922*, 10–11. See Robie, *For the Greatest Achievement*, for more on WW I aviation; Hallion, *Legacy of Flight*, for background on early twentieth century aviation; and Cartwright and Sprinkle, "History of Aeronautical Meteorology," for a brief discussion of aeronautical meteorology.

21. US Weather Bureau, *Report of the Chief, 1932–1933*, 5.

22. Weber, *Weather Bureau*, 40–41.

23. Yost, "Adjusting Rain Insurance Policies," 17–19.

24. US Weather Bureau, *Report of the Chief, 1922–1923*, 17; "Insurance Against Adverse Weather"; Palmer, "Weather Insurance," 67–70.

25. Weber, *Weather Bureau*, 36–37.

26. Ibid.

27. US Weather Bureau, *Report of the Chief, 1924–1925*, 3, 4.

28. See Weber, *Bureau of Chemistry and Soils* (appendixes 1, 2, and 6) and Milton Conover, *Office of Experiment Stations*.

29. US Weather Bureau, *Report of the Chief, 1932–1933*, 7; Marvin, "Committee on Research," 11; Kimball, "Recent Advances," 4.

30. Weber, *Weather Bureau*, 28.

31. Ibid., 32; US Weather Bureau, *Report of the Chief, 1921–1922*, 29.

32. Weber, *Weather Bureau*, 32; US Weather Bureau, *Report of the Chief, 1920–1921*, 22; US Weather Bureau, *Report of the Chief, 1924–1925*, 8.

33. "The Sun and the Weather," 25–26.

34. "Long-Range Weather Forecasting," 88.

35. Billings, "Is the Sun Fickle?" 269–271.

36. US Weather Bureau, *Report of the Chief, 1924–1925*, 8; DeVorkin, "Defending a Dream," 127.

37. "Calls Stars Weather 'Key.'"

38. Weber, *Weather Bureau*, 32; US Weather Bureau, *Report of the Chief, 1922–1923*, 22.

39. Weber, *Weather Bureau*, 19, 32–34.

40. Ibid., 35.

41. US Weather Bureau, *Report of the Chief, 1922–1923*, 5.

42. Conner, "Sunshine for Saturday," 105.

43. Moore, "Eight-Year Generating Cycle," 1–29; US Weather Bureau, *Report of the Chief, 1922–1923*, 5.

44. US Weather Bureau, *Report of the Chief, 1930–1931*, 4.

45. US Weather Bureau, *Report of the Chief, 1933–1934*, 2.

46. "Meteorological Data: Progress Report of Special Committee," 153–156. Letters, pro and con, appeared in the *Proceedings of the American Society of Civil Engineers* throughout 1933 under the title "Meteorological Data: Progress Report of Special Committee, Discussion."

47. RADM E. J. King to the Secretary of the Navy (H. L. Roosevelt, Acting), 8 July 1933 and 14 July 1933 (SAB, General 1933). For more on USS *Akron*, see Richard K. Smith, *Airships Akron and Macon*.

48. For an introduction to the literature on Wallace, see for example, Culver and Hyde, *American Dreamer*; Schapsmeier and Schapsmeier, *Henry A. Wallace*; and White and Maze, *Henry A. Wallace*. There has been little written on the inner workings of the Committee on the Weather Bureau. For more on the Science Advisory Board and its work during its short (1933–1935) existence, see Dupree, *Science in the Federal Government*, 350–358; Genuth, "Groping Towards Science Policy"; Kargon and Hodes, "Karl Compton"; Auerbach, "Scientists in the New Deal"; and Pursell, "Anatomy of a Failure."

49. Compton, "Report of the Science Advisory Board," 11. The other members were: W. W. Campbell, President, National Academy of Sciences; Isaiah Bowman, Chairman, National Research Council; Gano Dunn, President, J. G. White Engineering Corporation; Frank B. Jewett, Vice-President, American Telephone and Telegraph Company; Charles F. Kettering, Vice-President, General Motors Corporation; C. K. Leith, Professor of Geology, University of Wisconsin; John C. Merriam, President, Carnegie Institution; R. A. Millikan, Director, Norman Bridge Laboratory of Physics (Caltech).

50. Isaiah Bowman to Robert A. Millikan, 3 August 1933 (Millikan, 9/9.9).

51. Progress on the Committee on Weather Bureau, 4 December 1933; Karl T. Compton to Charles Dana Reed, 24 August 1933 (SAB, General 1933).

52. Isaiah Bowman, Minutes of SAB Committee on the Weather Bureau, 22, 25, and 26 August 1933 (SAB, General 1933).

53. Science Advisory Board, Minutes of the Meeting of the Committee on Meteorological Research, 24 September 1933; Science Advisory Board, Committee on Meteorological Research, Record of Meeting September 25, 1933 (SAB, General 1933).

54. Science Advisory Board, Committee on Meteorological Research, Record of Meeting September 25, 1933 (SAB, General 1933).

55. Bowman to Millikan, 2 October 1933 (SAB, General 1933).

56. Millikan to Wallace, 15 November 1933 (NAS, Agencies and Departments: Agriculture, WB 1933–1935); Compton, "Report of the Science Advisory Board," 53; US Weather Bureau, *Report of the Chief, 1933–1934*, 1; US Weather Bureau, *Report of the Chief, 1934–1935*, 1, 7.

57. US Weather Bureau, *Report of the Chief, 1920–1921*, 20.

58. US Weather Bureau, *Report of the Chief, 1922–1923*, 10; US Weather Bureau, *Report of the Chief, 1923–1924*, 18; Brooks, "*Monthly Weather Review* Halved," 154; S. D. F., "News Item," 212.

59. From Science Service: "Nine and a Half Million Cut from Government Research." See also Dupree, *Science in the Federal Government*, 343–350.

60. "Proceedings of the First Meeting," 13; "Some Weather Bureau Projects"; Marvin quoted in US Weather Bureau, *Report of the Chief, 1919–1920*; "Efficiency of the Weather Bureau Endangered"; Brooks, "Reclassification," 163–164; Weber, *Weather Bureau*, 49, 70; Milton Conover, *Office of Experiment Stations*, 128, 166; Weber, *Bureau of Chemistry and Soils*, 165, 188.

61. "Increased Pay for Weather Bureau Employees"; "Personnel of the Weather Bureau."

62. Brooks, "Many Government Meteorologists Retired."

63. S. D. F., "Retrenchment"; US Weather Bureau, *Report of the Chief, 1932–1933*, 3; "Effect of Economy Program on U.S. Weather Bureau"; "Weather Bureau Salaries."

64. Fassig, "Weather Bureau."

65. Byers, "Founding of the Institute of Meteorology," 1343.

66. Quoted in Nelson, "Aërology in the Navy," 524.

67. Keyser, "Aerological Work."

68. For more information on the aerological training of naval officers during World War I, see John Conover, *Blue Hill*, 144–149; Nelson, "Aërology in the Navy," 523; Bates and Fuller, *Weather Warriors*, 23–29.

69. Keyser, "From the Committees"; "Naval Officers' Advanced Course in Meteorology at the Weather Bureau, Washington, D.C."

70. Nelson, "Aërology in the Navy," 525; Bates and Fuller, *Weather Warriors*, 37–38.

71. Bates and Fuller, *Weather Warriors*, 37.

72. Namias, "Francis W. Reichelderfer"; O'Brien, "Comments"; Reichelderfer, "Postgraduate Course."

73. Hallion, *Legacy of Flight*, 219. Carl-Gustav Rossby's first name appears as "Carl-Gustav" in articles he wrote and as "Carl-Gustaf" in articles written about him. I have used "Carl-Gustav" throughout the text.

74. Byers, "Carl-Gustaf Arvid Rossby."

75. Reichelderfer, "Postgraduate Course."

76. Reichelderfer, "Polar Front Theory"; Bates and Fuller, *Weather Warriors*, 38.

77. Reichelderfer, "Polar Front Theory"; Taba, *Bulletin Interviews*, 91; 1931 Trip Reports (Reichelderfer, 5/1); Reichelderfer, "Atmospheric Sciences."

78. O'Brien, "Navy's Part," 390.

79. Hale, "Navy's Part"; O'Brien, "Navy's Part," 385, 392.

80. For details of the Army's control of weather services and their subsequent shift to civilian control, see Whitnah, *History of the United States Weather Bureau*, 22–60; Larew, *Meteorology in the U.S. Army Signal Corps*, 1–22; Fleming, *Meteorology in America*, 157–166; Bates and Fuller, *Weather Warriors*, 9–14; Fuller, *Thor's Legions*, 3–7; and Dupree, *Science in the Federal Government*, 187–192.

81. Flammer, "Meteorology in the United States Army," 11–15; on sound-ranging see Kevles, *The Physicists*, 126–131.

82. Bates and Fuller, *Weather Warriors*, 18.

83. Flammer, "Meteorology in the United States Army," 18.

84. Reed, "Papers Presented"; Wobbe, "Meteorology in its Application to Artillery Fire."

85. Flammer, "Meteorology in the United States Army," 33–34.

86. "From the Committees"; Fuller, *Thor's Legions*, 18–19.

87. Larew, *Meteorology in the U.S. Army Signal Corps*, 39–41.

88. Ibid., 42; Flammer, "Meteorology in the United States Army," 48.

89. Fuller, *Thor's Legions*, 18–19.

90. Van der Linden, *Airlines and Air Mail*, 260–277; Flammer, "Meteorology in the United States Army," 48. My thanks to an anonymous reviewer for the Van der Linden citation.

91. Testimony of Colonel Clark, quoted in memorandum to Edward L. Bowles, 17 August 1943 (Bowles, 30/3).

92. Bates and Fuller, *Weather Warriors*, 48.

93. Terrett, *Signal Corps*, 28, 32; Brophy et al., *Chemical Warfare Service*, 32.

Notes to Chapter 2

1. Unpublished table of the National Research Council, Table III, Doctorates Conferred According to Subjects (1923) (NAS, Research Information Service, 1920–1923, Information Files Doctorates Conferred).

2. Ward, "Meteorological Instruction," 554.

3. Ibid. For contemporary assessments of the state of academic meteorology in the early twentieth century, see Waldo, "Study of Meteorology"; Abbe, "Progress of Science"; "Curiosities of Science and Invention: Meteorology in American Universities," 343; and Reed, "Meteorological Observations."

4. Brooks, "Collegiate Instruction."

5. Fassig, "Signal Corps School."

6. Brooks, "Collegiate Instruction."

7. Kevles, *Physicists*, 118–119, 132–133, 138.

8. Fassig, "Signal Corps School," 561.

9. Shaw, "Outlook."

10. Brooks, "General Extent."

11. Quoted in Koelsch, "From Geo- to Physical Science," 526.

12. Brooks, "Meteorological Instruction." The committee membership sheds light on the issue of meteorological offerings by identifying affiliations: W. M. Wilson, Department of Meteorology, Cornell University *and* Section Director, Weather Bureau Office, Ithaca, NY; W. I. Milham, Department of Astronomy, Williams College; H. E. Simpson, Department of Geology, University of North Dakota; F. L. West, Utah Agricultural College.

13. J. Smith, "Agricultural Meteorology."

14. Brooks, "Meteorological Instruction."

15. "Notes of Interest to Teachers."

16. "Meteorology at the Southern Branch of the University of California."

17. "Proceedings of the First Annual Meeting: Report of Committee on Meteorological Instruction."

18. Palmer, "Miscellaneous Notes."

19. "The Second Year of the Society."

20. "University News."

21. Shipman, "Meteorology and Climatology."

22. Humphreys, "Claims of Meteorology."

23. "Notes."

24. Jacobs, "Survey of Instruction," 11, 26, 32, 39.

25. Ibid., 49, 50, 95.

26. Harding, "Meteorology and Climatology."

27. Spilhaus, well known for his invention of the bathythermograph for measuring the temperature of water with depth, left NYU in 1948 to become the Dean of the Institute of Technology at the University of Minnesota.

28. Stone, "New Department"; Woodman, "New York University."

29. Meisinger, "American Doctors of Meteorology."

30. "Four New Doctors of Philosophy in Meteorology or Climatology."

31. "Personal Notes."

32. Stone, "New Department."

33. Koelsch, "From Geo- to Physical Science," 522.

34. Kimball, "Review of Recent Advances," 187.

35. "Fog—A Method of Prevention?"

36. Bean, "Weather and Crop Research."

37. Koelsch, "From Geo- to Physical Science," 526. For more details of the meteorology program at Caltech, see Lewis, "Cal Tech's Program."

38. John Conover, *Blue Hill,* 135, 156, 157, 163, 168, 172.

39. "The American Meteorological Society."

40. Quoted in Brooks, "Our Society's First Decade," on 10.

41. Ibid.

42. "Committees."

43. "From the Committees: Research," 17–18.

44. "From the Committees: Meteorological Instruction," 16.

45. Talman, "Public Information."

46. On Huntington's views about climate and environmental determinism, see Livingstone, *Geographical Tradition*, 230–231.

47. Huntington, "Physiological Meteorology." Huntington's ideas were widely published in the popular press. For example, see Barton, "What the Weather Does to You"; Huntington, "New Astrology" and "What the Weather Does to Us." There has been very little historical research done on medical meteorology. For an introduction, see Feldman, "Late Enlightenment Meteorology"; Genevieve Miller, "Airs, Waters, and Places"; Riley, *Campaign Against Disease*; and Sargent, *Hippocratic Heritage*.

48. Woolard, "Mathematical Meteorology"; Richardson, *Weather Prediction*, 219. Richardson, a Quaker pacifist, had written his work while serving with the Friends Ambulance Unit during World War I.

49. Marvin, "Status, Scope and Problems."

50. Woolard, "Extratropical Cyclone."

51. Woolard, "Theoretical Meteorology," 78–81.

52. Brooks, "Outline."

53. Beals, "Possible to Predict."

54. Blochman, "Long-Range Forecasting."

55. Dupree, *Science in the Federal Government*, 364.

56. Willett, "Importance of Observations."

57. For an introduction to the history of air-sea interaction, see Charnock, "Ocean-Atmosphere Interactions."

58. Byers and Page, "Conference on Long-Range Forecasting."

59. Woolard, "Conference on Long-Range Forecasting." An isentropic chart is a synoptic chart with plotted meteorological elements (pressure, wind, temperature, moisture) on a surface of constant potential temperature. Potential temperature is the temperature that an unsaturated parcel of dry air would have if brought adiabatically (i.e., without exchanging heat with its environment) from its initial state to a standard pressure of 1,000 millibars (or 100 kilopascals).

60. Talman, "Ice Coating."

61. McAdie, "Hazard of Sub-Cooled Fog."

62. Lankford, *American Astronomy*, 362. There were 362 professional astronomers in 1940.

63. Kevles, *Physicists*, 202.

Notes to Chapter 3

1. Gregg, "Introductory Remarks."

2. Telegram from Albert L. Barrows to Millikan, 15 September 1938; Millikan to Frank R. Lillie and Ross G. Harrison, 15 September 1938 (NAS/SWB).

3. Barrows to Lillie, 16 September 1938 (NAS/SWB).

4. Millikan to Weather Bureau Advisory Committee (Isaiah Bowman, Karl Compton, H. D. Hughes, and J. B. Lippincott), 5 October 1939 (NAS/SWB).

5. Record of telephone call: Barrows and Compton, 18 October 1938 (NAS/SWB).

6. Compton to Wallace, 24 October 1938; Leahy to Millikan, 20 October 1938 (NAS/SWB).

7. Whitnah, *History of the United States Weather Bureau*, 132.

8. Reichelderfer to Wallace, 15 November 1938 (Reichelderfer, 4/9).

9. Willett, "Rossby."

10. Report of the Subcommittee on the Weather Bureau, April 1939 (NAS/SWB).

11. Reichelderfer to Millikan, 28 November 1938 (Reichelderfer, 4/9).

12. "Expanding Instruction in Meteorology and Climatology"; Byers, "Carl-Gustaf Rossby, the Organizer," 59.

13. F. W. Reichelderfer, "The Weather Bureau Program for 1939," a talk presented before the Atlanta Meeting of the American Meteorological Society, 21 April 1939 (Reichelderfer, 1/7).

14. "University Training in Meteorology under Civil Aeronautics Act."

15. Fuller, *Thor's Legions*, 32–33.

16. "Meteorological Education in the United States."

17. "Announcements."

18. Byers, "Founding of the Institute of Meteorology," 1343.

19. Ibid.

20. Byers, "Carl-Gustaf Rossby, the Organizer," 58.

21. "Announcements."

22. The effect of rapid demobilization on scientific research efforts connected with World War I are briefly discussed in Kevles, *Physicists,* 147, and Dupree, *Science in the Federal Government,* 324–325.

23. Yates, "Remarks."

24. Bates and Fuller, *Weather Warriors,* 52.

25. Charles F. Sarle, 15 January 1942 (Wexler, 1/Gen. Corr. 1942).

26. Kaplan to Smith, 10 Nov 1942; Philemon Church to Lee Paul Sieg, 11 Nov 1942 (WU President, 110/Met Training).

27. Church to Sieg, 11 November 1942 (WU President, 110/Met Training). Whitnah, *History of the United States Weather Bureau* (p. 21) reports that the bureau had 1,494 full-time employees in 1938; 3,218 in 1944; and 4,727 in 1946. By late 1944, over 700 had left for active military service. According to O'Brien and Grice, "Women in the Weather Bureau," there were only two women listed as either observer or forecaster in 1941. Office staffs were all men. By 1945, 900 women worked for the bureau as either clerks or junior observers; most were temporary employees.

28. Rossby to Col. D. Z. Zimmerman, 28 September 1942 (UMC, 1).

29. Kaplan to Sieg, 27 October 1942 (WU President, 110/Met Training).

30. Rossby to Sieg, 3 November 1942 (WU President, 110/Met Training).

31. There were eleven Pre-meteorological Centers: Brown University, MIT, NYU, State University of Iowa, University of California (Berkeley), University of Michigan, University of Minnesota, University of New Mexico, University of North Carolina, University of Washington, University of Wisconsin. There were twelve Basic Pre-meteorological Centers: Amherst College, Bowdoin College, Carleton College, Denison University, Hamilton College, Haverford College, Kenyon College, Pomona College, Reed College, University of Chicago, Vanderbilt University, and University of Virginia.

32. Sieg to Rossby, 5 November 1942 (WU President, 110/Met Training). Ultimately, the University of Washington and other schools hired high school mathematics teachers to instruct lower level courses.

33. Rossby to Sieg, 7 November 1942 (WU President, 110/Met Training).

34. Rossby to Sieg, 10 November 1942 (WU President, 110/Met Training).

35. Bates and Fuller, *Weather Warriors,* 57.

36. Rossby to Major General B. M. Giles, 10 September 1943 (Bowles, 30/4).

37. F. W. Reichelderfer, Chief, USWB, "Summary for Secretary's [Commerce] Report, Fiscal Year 1942," ca. summer 1942 (Reichelderfer, 7/10).

38. Koelsch, "Geo- to Physical Science," 531.

39. Bates and Fuller, *Weather Warriors*, 51. In weather typing, current weather conditions are matched to ones that have occurred in past years. Once a good match has been found, the subsequent maps are examined and used as forecasts for the next few days.

40. Koelsch, "Geo- to Physical Science," 530.

41. Rossby to Wallace, 6 October 1942 (UMC, 4/Reichelderfer).

42. Rossby to Sieg, 3 November 1942 (WU President, 110/Met Training).

43. A. F. Carpenter, University of Washington representative at the Conference on Army-Sponsored Meteorological Training, University of Chicago, 18–19 December 1942 (WU President, 110/Met Training).

44. Report of the University of Washington representative at the conference of representatives of colleges participating in the Army-Sponsored Meteorology Training Programs, University of Chicago, 8–9 January 1943 (WU President, 110/Met Training).

45. Rossby to Edward Bowles, 18 April 1943 referenced in UMC Meeting Minutes of 4 June 43 (Wexler, 2/1943).

46. Ibid.

47. For the role of physicists during World War Two see Kevles, *Physicists*, chapters 19–21.

48. Henry L. Stimson to Robert Maynard Hutchins, 1 May 1943 (Bowles, 30/2).

49. Bowles to Rossby, 30 April 1943 (Bowles, 32/1).

50. Rossby to Houghton, 25 May 1943 (Bowles, 40/2); Bowles to Secretary of War, 23 August 1943 (Bowles, 30/3).

51. UMC Meeting minutes, 3–6 June 1943 (Wexler, 2/1943).

52. Brophy and Fisher, *Chemical Warfare Service*, 356.

53. UMC Meeting minutes, 3–6 June 1943 (Wexler, 2/1943).

54. Research Projects prepared for JMC Research Committee Conference (Army, Navy and Weather Bureau cooperating), 23 January 1943 (UMC, 4/Reichelderfer).

55. Thomas B. Marshall to Colonel Thompson, 26 Aug 1943 (UMC, 7/Research).

56. UMC Meeting Minutes, p. 19, 6 December 1943 (UMC, 2/Blue Folder).

57. Dupree, *Science in the Federal Government*, 369–371.

58. UMC Meeting Minutes, p. 24, 6 December 1943 (UMC, 2/Blue Folder).

59. Jörgen Holmboe (UCLA), UMC Meeting Minutes, 25, 6 December 1943 (UMC, 2/Blue Folder).

60. "Secretary's Report."

61. UMC Meeting Minutes, 26–27, 6 December 1943 (UMC, 2/Blue Folder).

62. Ibid., 33, 43, 44; Summary of the Forecaster's Conference, 15–20 May 1944, issued 1 August 1944 (USWB, 2/Multiple Address Letters).

63. UMC Meeting Minutes, pp. 26–27, 6 December 1943 (UMC, 2/Blue Folder).

64. Ibid., 40, 58.

65. Ibid., 26–27.

66. Ibid.

67. Ibid., 46–47.

68. Ibid., 48.

69. Ibid., 25.

70. C. F. Brooks to Wexler, 8 August 1944 (Wexler, 2/1944).

71. Washington, "Foreword," vi.

72. UMC Meeting Minutes, p. 70, 6 December 1943 (UMC, 2/Blue Folder).

73. Church to Sieg, 11 November 1942 (WU President, 110/Met Training).

74. Rossby, "Message to Members," 268–269.

75. See Good, "Rockefeller Foundation," and Doel, "Earth Sciences."

76. Byers, Kaplan, and Minser, "Teaching of Meteorology," 95.

77. Ibid.

78. Rossby, "Message to Members," 268–269.

79. Quoted in Malone, "Atmospheric Sciences," 218.

80. Spengler, "From the Executive Secretary," 255.

81. Quoted in Yates, "Remarks."

82. Byers, "Recollections," 217.

Notes to Chapter 4

1. See, e.g., Williams, *History of Computing Technology*; Campbell-Kelly and Aspray, *Computer*; and Ceruzzi, *Modern Computing*.

2. On von Neumann, see Dieudonne, "John von Neumann" and Aspray, *John von Neumann.*

3. See Weart, *Nuclear Fear.*

4. See, e.g., Hirt, *Conspiracy of Optimism*; Scott, *Seeing Like a State*; and White, *Organic Machine.*

5. Weather control has been a dream of mankind since the earliest days of appealing to the gods for relief from undesirable weather. The few historical treatments of early weather modification efforts in the United States include Spence, *Rainmakers*, and Townsend, *Making Rain.*

6. This story, as told in chapter 2 of Aspray, *John von Neumann*, and chapter 10 of Nebeker, *Calculating the Weather,* was almost completely about von Neumann. The archival record shows that meteorologists located a considerable distance from Princeton, in particular Rossby and Reichelderfer, were crucial to the success of "von Neumann's" Meteorology Project.

7. See, e.g., Wang, "Liberals"; Sapolsky, *Science and the Navy*; Forman, "Behind Quantum Electronics"; Owens, "Science in the United States"; Dennis, "Historiography of Science"; and Lowen, *Cold War University.*

8. Byers, 29 February 1944, UMC Executive Meeting Minutes, iii (UMC, 1/UMC Meeting at USWB).

9. Houghton, ibid., iv.

10. H. J. Stewart, ibid., iv.

11. Byers, ibid., iv.

12. Houghton, ibid., iv.

13. After 1945, scientific disciplines responded to the availability of government funding in different ways. Physics, for example, expected to receive significant government funds, while astronomers still depended upon private patrons. See Doel, *Solar System Astronomy*; DeVorkin, "Organizing for Space Research"; and Appel, *Shaping Biology.*

14. Gutenberg, Kaplan, Byers, 29 February 1944, UMC Executive Meeting Minutes, v (UMC, 1/UMC Meeting at USWB).

15. "The Weather Bureau Questionnaire on Research Needs." Orography, a branch of physical geography, deals with mountains.

16. See Aspray, *John von Neumann*, chapter 2, for a short discussion of ENIAC. For a longer treatment of ENIAC, other Eckert-Mauchly computers, and the roles of Eckert and Mauchly in computing, see Stern, *ENIAC to UNIVAC*. McCartney, *ENIAC*, gives a more popular account of ENIAC's development and the personalities that played a role in it. Contemporary descriptions include Goldstine and Goldstine, "Electronic Numerical Integrator and Computer," and Berkeley, *Giant Brains*, chapter 7.

17. John W. Mauchly, "Note on Possible Meteorological Use of High Speed Sorting and Computing Devices," 14 April 1945, copied by F. W. Reichelderfer on 24 January 1946 and marked "Confidential." (Wexler, 2/1945). What Reichelderfer might have thought about Mauchly's ideas in spring 1945 is unknown. According to Mauchly's notes, he did not visit with Reichelderfer.

18. Wexler to Reichelderfer, 18 October 1946 (Wexler, 2/1946). Weather Bureau personnel actually considered this idea, albeit very briefly, before rejecting it.

19. Reichelderfer to Zworykin, 4 December 1945 (Wexler, 2/1945).

20. Reichelderfer notation on letter from E. U. Condon to Reichelderfer, 26 November 1945 (Wexler, 2/1946).

21. Reichelderfer to Zworykin, 4 December 1945 (Wexler, 2/1945).

22. Reichelderfer to Sarle, 4 December 1945 (Wexler, 2/1946).

23. Reichelderfer to Weather Bureau staff, 29 December 1945 (Wexler, 2/1945). Original emphasis.

24. Reichelderfer to von Neumann, 29 December 1945 (Wexler, 2/1946).

25. Frank Aydelotte to von Neumann, 9 November 1945 (von Neumann, 12/1).

26. In a speech given in New York in January 1961, then Rear Admiral de Florez called for a $100 million/year program of meteorological research with the ultimate aim of weather modification and control. (Wexler, 35/Weather Modification–4).

27. Shallet, "Weather Forecasting."

28. Reichelderfer to Zworykin, 11 January 1946 (Wexler, 2/1946).

29. Record of telephone conversation between Reichelderfer and Col. Gillen, 21 January 1946 (Wexler, 2/1946).

30. Zworykin to Reichelderfer, 14 January 1946 (Wexler, 2/1946).

31. Wexler and Jerome Namias to Reichelderfer, 8 February 1946 (Wexler, 2/1946). Namias, who had worked under Rossby at MIT on Bankhead-Jones Act-funded long-range weather forecasting techniques, continued that work throughout his career at the Weather Bureau. A childhood friend of Wexler's, he was also his brother-in-law. Army Air Force meteorologist and mathematician Gilbert Hunt also attended this meeting.

32. Wexler and Namias to Reichelderfer, 26 February 1946 (Wexler, 2/1946).

33. Reichelderfer to the Secretary of Commerce (Henry A. Wallace), 18 February 1946 (Wexler, 2/1946). The Weather Bureau had formerly been under the Department of Agriculture.

34. Gilbert Hunt to Wexler, 18 March 1946 (Wexler, 2/1946).

35. Richardson, *Weather Prediction*. For a short discussion of Richardson and his numerical modeling attempt, see Hayes, "Weatherman." For a look back at Richardson's method, see Chapman, "Introduction." For a discussion of Richardson's eclectic interests—a Quaker, he also studied conflicts—see Ashford, *Prophet—or Professor?* For a reconstruction of Richardson's forecast see Lynch, "Weather Forecasting."

36. Wexler's notes dated 12 April 1946 (Wexler, 2/1946). Von Neumann suggested that a meteorology department be established at Princeton University, but that never happened. Whether it would have helped the project could be debated. With many academic meteorologists skeptical of the usefulness of the computer, it may have been a hindrance. Since there were not enough academic meteorologists to go around anyway, it is not clear where they might have found the personnel to staff it. The dearth of practicing meteorologists in the postwar years, despite the thousands trained during the war, is an interesting topic that needs to be explored.

37. Von Neumann to Rossby, 6 February 1946 (von Neumann, 15/7).

38. Rossby to Reichelderfer, 16 April 1946 (von Neumann, 15/7).

39. Ibid.

40. Reichelderfer to Rossby, 24 April 1946 (Wexler, 2/1946).

41. Rossby to LCDR D. F. Rex, 23 April 1946 (Wexler, 2/1946). Rex, a Navy Aerologist (as they were then called), later earned his PhD in meteorology under Rossby in Stockholm. In the late 1940s, Rossby would return to Sweden to establish a Department of Meteorology at the University of Stockholm.

42. Rossby to von Neumann, 23 April 1946 (Wexler, 2/1946).

43. Frank Aydelotte to LCDR D. F. Rex, 8 May 1946 (von Neumann, 15/6). A definition of terms is in order here. The stratosphere (the second "layer" in the atmosphere) extends from the top of the troposphere (approximately 10–17 km above the earth's surface) to the bottom of the mesosphere (approximately 50 km above the earth's surface). The troposphere occupies the space between the earth's surface and the bottom of the stratosphere. The "polar front" is the semi-permanent front separating tropical and polar origin air masses. "Cyclones" are large-scale regions of low pressure that turn in a counter-clockwise rotation in the northern hemisphere (clockwise in the southern hemisphere)—in meteorological usage they are not tornadoes or similar small-scale circulations. A system is "stable" if small disturbances have only small effects; it is "unstable" if a small disturbance generates a large effect.

44. Justification Memorandum, PD #EN1-22/00028, The Institute for Advanced Study, 6 June 1946 (von Neumann, 15/6). The idea of "control" over the weather is a recurrent one throughout the project, but it generally comes from those who are not meteorologists: Zworykin and von Neumann. During this same period Nobel Laureate Irving Langmuir, his assistant Vincent Schaefer, and Bernard Vonnegut, actively worked on weather control through seeding. Vonnegut had the most meteorological

background. Most meteorologists were struggling to understand the atmosphere and get a prediction out for the next day—control of the general circulation was not an issue for them even though work on smudging orchards to keep fruit from freezing was. The theme of control over nature and its manifestation in weather modification for agriculture, military or diplomatic reasons recurs throughout the Cold War and needs a close examination.

45. Frank Aydelotte to LCDR D. F. Rex, 8 May 1946 (von Neumann, 15/6).

46. Raymond P. Montgomery to Wexler, 14 May 1946 (Wexler, 2/1946).

47. Haurwitz to Wexler, 21 June 1946 (Wexler, 2/1946).

48. Wexler to Hunt, 3 May 1946 (Wexler, 2/1946).

49. In his efforts to get Paul Queney on the team, Rossby asked that the University of Algiers extend the leave of absence that had allowed Queney to spend time at the University of Chicago. Rossby to Pierre Auger, Directeur de l'Enseignement Superieur, Ministère de l'Education National, Paris, 11 May 1946 (von Neumann, 15/7).

50. Von Neumann to Queney, 18 May 1946; von Neumann to Haurwitz, 6 June 1946; von Neumann to Wexler, 14 June 1946 and 29 June 1946 (all from Wexler, 2/1946).

51. Sverdrup to Reichelderfer, 2 June 1946 (Wexler, 2/1946).

52. Hans Panofsky to Wexler, 22 July 1946 (Wexler, 2/1946).

53. Haurwitz to Reichelderfer, 14 May 1946 (Wexler, 2/1946).

54. Robert D. Elliott to Reichelderfer, 4 June 1946 (Wexler, 2/1946).

55. Von Neumann to Jule Charney, 14 August 1946 (Charney, 16/516).

56. Minutes, Conference on Meteorology, August 29–30, 1946 (Charney, 4/134).

57. Wexler to Reichelderfer, 12 September 1946 (Wexler, 2/1946).

58. Wexler to Reichelderfer, 18 October 1946 (Wexler, 2/1946).

59. Wexler to Reichelderfer, 4 November 1946 (Wexler, 2/1946).

60. Regis, *Einstein's Office*, 111. Local Princeton residents were concerned that the new computer would make a lot of noise, and wanted it well away from Princeton's residential districts.

61. Aspray, *John von Neumann*, 58–59.

62. Wexler to Reichelderfer, 4 November 1946 (Wexler, 2/1946).

63. Progress Report for the period of July 1, 1946 to November 15, 1946 on Contract N6ori-139, Task I (Meteorology Project) (von Neumann, 15/16).

64. Wexler to Reichelderfer, 15 November 1946 (Wexler, 2/1946). Visiting von Neumann were Athelstan Spilhaus and Hans Panofsky (both of NYU), Walter Munk (Scripps Institution of Oceanography), and Henry Stommel (Woods Hole Oceanographic Institution).

65. Wexler to Reichelderfer, 22 November 1946 (Wexler, 2/1946).

66. Notes on the Chicago Conference on Problems of Meteorological Research, December 9–13, 1946 prepared by S. Hess (Wexler, 3/1947-1).

67. Meeting for Discussion of Meteorological Problems, University of Chicago, December 9–13, 1946—report of AAF Lt. Philip D. Thompson (Thompson). N.B.: Since the Philip D. Thompson collection was being processed when I used it, no box and folder information is provided in these notes.

68. Headquarters, North Atlantic Division, Air Transport Command to Commanding General, Air Transport Command, 30 October 1944 (Thompson).

69. Headquarters, Eighth Weather Squadron to Station Weather Officer, 1388th AAF Base Unit, 22 November 1944 (Thompson).

70. Thompson to Commanding Officer, Headquarters, Army Air Forces Weather Service, 17 December 1945 (Thompson).

71. A thorough examination of the *New York Times* index revealed no interview with Zworykin or von Neumann.

72. Thompson, "Notes," 757–758.

73. History of the 72nd AAFBU Detachment, IAS, November 22, 1946—December 31, 1946 (Classification: Restricted) by Philip D. Thompson (Thompson).

74. Ibid.

75. Thompson to Wexler, 14 January 1947; Report of Temporary Duty, AAF Lt. Philip D. Thompson, 20 January 1947 (Thompson).

76. Charney, "Dynamics of Long Waves."

77. Charney to Thompson, 7 February 1947 (Thompson).

78. Charney to Rossby, ca. 19 March 1947 (Charney, 14/460).

79. Philip D. Thompson, "Survey of the Electronic Computer Project at the Institute for Advanced Study," 6 March 1947; Thompson to Delmar Crowson, 11 March 1947 (Thompson).

80. Thompson to Robert Bundgaard, 27 March 1947 (Thompson).

81. Wexler to Thompson, 17 March 1947 (Wexler, 2/1947-5).

82. Thompson to Wexler, 21 April 1947 (Thompson).

83. Reichelderfer to Wexler, 2 June 1947 (Wexler, 3/1947-4).

84. Cahn to Wexler, 12 June 1947 (Wexler, 3/1947-7).

85. Philip D. Thompson, "Report of Progress, September 1947" (Thompson).

86. Wexler to Reichelderfer, 21 November 1947 (Wexler, 3/1947-6).

87. Philip D. Thompson, "Concerning the Numerical Forecasting Project," ca. 1947 (Thompson).

88. Charney to Thompson, 4 November 1947 (Thompson).

89. Charney to Jacob Bjerknes, 14 January 1948 (Charney, 4/120).

90. Von Neumann to Charney, 19 November 1947 (von Neumann, 15/1).

91. Charney to von Neumann, 2 January 1948 (von Neumann, 15/1). See Petterssen's *Weathering the Storm*, for his account of his role in the D-Day invasion forecast.

92. Von Neumann to Charney, 6 February 1948 (von Neumann, 15/1).

93. Charney to von Neumann, 29 February 1948 (Charney, 16/517).

94. Jacob Bjerknes to Charney, 4 March 1948 (Charney, 4/120).

95. Thompson to Joe Fletcher, 8 March 1948 (Thompson).

96. Wexler to Reichelderfer, 21 November 1947 (Wexler, 3/1947-6).

97. Wexler to Reichelderfer, 12 May 1948 (Wexler, 3/1948).

98. Report of Progress, ONR Meteorology Project, IAS, Princeton, December 15, 1947—May 15, 1948, 18 May 1948 (Charney, 9/304).

Notes to Chapter 5

1. Rossby to Charney, 28 October 1948 (Charney, 14/460).

2. Rossby to Charney, 9 January 1949 (Charney, 14/460).

3. Charney to Rossby, 19 March 1947 (Charney, 14/460).

4. Charney to Rossby, 15 September 1948 (Charney, 14/460).

5. Rossby to Charney, 25 September 1948 (Charney, 14/460).

6. Rossby to Charney, 7 January 1951 (Charney, 14/459).

7. For details of the attributes of research schools, see Geison, "Research Schools." Drawing on the work of J. B. Morrell, Geison described 14 attributes of a research school. Among them was a charismatic leader who possesses a research reputation, an "informal" leadership style, and institutional power. Furthermore, the research school produces a significant number of students and then finds them positions, effectively

spreading the school's influence. Rossby and his many acolytes who played key roles in the development of numerical weather prediction embodied the attributes of a successful research school.

8. Thompson to Paul Worthman, 1 March 1949 (Thompson).

9. Byers, "Carl-Gustaf Rossby, the Organizer."

10. See Phillips, "Carl-Gustaf Rossby," and Lewis, "Carl-Gustaf Rossby," for discussions of Rossby as a meteorological leader and mentor.

11. Rossby, "Propagation of Frequencies."

12. Progress report of the Meteorology Group at the IAS, 1 July 1948—30 June 1949 (Charney, 9/304).

13. In fact, it would later become known that models would have to resolve the gravity waves during the whole integration.

14. Glickman, *Glossary*, 76.

15. Charney to von Neumann, 24 August 1948 (von Neumann, 15/1).

16. Von Neumann to Charney, 17 September 1948; Charney to von Neumann, 21 September 1948 (Charney, 16/517).

17. Charney to Rossby, 15 September 1948 (Charney, 14/460).

18. Rossby to Charney, 25 September 1948 (Charney, 14/460); Yeh, "Energy Dispersion."

19. Rossby to Charney, 27 September 1948 (Charney, 14/460).

20. Rossby to Charney, 24 October 1948 (Charney, 14/460).

21. Rossby to Charney, 28 October 1948 (Charney, 14/460).

22. Rossby to Charney, 13 December 1948 (Charney, 14/460).

23. Charney to Rossby, 20 December 1948 (Charney, 14/460).

24. This method was ultimately published in Charney and Eliassen, "Numerical Method."

25. Rossby to Charney, 9 January 1949 (Charney, 14/460).

26. Rossby to Charney, 4 February 1949 (Charney, 14/460).

27. Rossby to Charney, undated, ca. late February 1949 (Charney, 14/459).

28. Rossby to Charney, 5 April 1949 (Charney, 14/459).

29. Rossby to Charney, 27 March 1949 (Charney, 14/459).

30. Rossby to Charney, 23 April 1949 (Charney, 14/459).

31. Ibid. The paper is Charney and Eliassen, "Numerical Method."

32. Participating in the meeting were Sutcliffe (UK), van Mieghem (Belgium), Godske and Høiland (Norway), Lysgaard and Andersen (Denmark), Palmén and Keräuen (Finland), Faegn (Norway - botanist), Ahlmann (Sweden), Kuo (China), Rex and Hutchinson (US Navy officers), Namias (USWB), and Reuter (Austria).

33. Rossby to Charney, 8 May 1949 (Charney, 14/459).

34. Rossby to Platzman, 8 May 1949 (Charney, 14/460).

35. Rossby to Charney, 19 May 1949 (Charney, 14/459).

36. Progress Report of the Meteorology Group at the IAS, 1 July 1948–30 June 1949 (Charney, 9/304).

37. For a Rossby (or planetary) wave, the wave speed c is given by $c = U - \beta L^2/4\pi^2$, where U is the mean westerly flow, β is the Rossby parameter, and L is the wavelength. See Platzman, "Rossby Wave," for a discussion of Rossby waves in the atmosphere and ocean.

38. Charney to Rossby, 24 May 1949 (Charney, 14/459).

39. Charney to Jacob Bjerknes, 21 April 1949 (Charney, 4/120).

40. Charney to Rossby, 24 May 1949 (Charney, 14/459).

41. Von Neumann to Rossby, 13 June 1949 (Charney, 14/459).

42. Charney to Rossby, 15 June 1949 (Charney, 14/459).

43. Alan T. Waterman (ONR) to von Neumann, 17 June 1949 (Charney, 14/459).

44. Joseph Smagorinsky to Charney and Eliassen, 28 July 1949 (Charney, 14/473).

45. Reichelderfer to MGEN E. S. Hughes, 22 September 1949 (von Neumann, 15/2).

46. MGEN H. B. Sayler to Reichelderfer, 29 September 1949 (von Neumann, 15/2).

47. Reichelderfer to von Neumann, undated, ca. early October 1949 (von Neumann, 15/2).

48. Von Neumann to Reichelderfer, 5 October 1949 (von Neumann, 15/2).

49. Both Aspray, *John von Neumann* (142), and Nebeker, *Calculating the Weather* (146), credit von Neumann with making the arrangements to use ENIAC. The letters from the von Neumann collection show that Reichelderfer obtained permission to use ENIAC. Von Neumann dealt with Aberdeen on using the ENIAC for the Meteorology Project's initial run only after Reichelderfer had cleared the way.

50. Rossby to Platzman and Charney, 9 October 1949 (Charney, 14/459). Palmén, who had known Rossby since 1930 when they met in Bergen, joined Rossby in Chicago for

2 years, starting in 1946. Once Rossby left for Stockholm, he still visited Chicago to work for extended periods of time.

51. Joseph Smagorinsky to Charney, 24 October 1949 (Charney, 14/473).

52. Charney to Joseph Smagorinsky, 5 December 1949 (Charney, 14/473).

53. Joseph Smagorinsky to Charney, 8 December 1946 (Charney, 14/473).

54. Charney to Rossby, 21 December 1949 (Charney, 14/459).

55. Charney to Rossby, 21 December 1949 (Charney, 14/459).

56. Von Neumann to E. T. Eady, 9 January 1950 (Charney, 16/517).

57. The Institute for Advanced Study, The Meteorology Group, Progress Report July 1, 1949–June 30, 1950, Contract No. N-6-ori-139 (Charney, 9/304).

58. Charney to Rossby, 21 February 1950 (Charney, 14/459).

59. Rossby to Charney, 23 February 1950 (Charney, 14/459).

60. Platzman, "ENIAC Computations." According to Platzman, Rossby was present for part of one day during the entire five-week run.

61. Reichelderfer to Col. Alden P. Taber, 10 April 1950 (von Neumann, 15/2).

62. The results were published a year later in Charney, Fjørtoft, and von Neumann, "Numerical Integration."

63. Charney to Rossby, 5 June 1950 (Charney, 14/459).

64. Bolin went on to have a distinguished career in geophysics, both in Sweden and internationally, after assuming the leadership of the International Meteorological Institute in Stockholm following Rossby's death in 1957.

65. Rossby to Charney, 13 June 1950 (Charney, 14/459).

66. Platzman to Charney, 18 June 1950 (Charney, 14/451).

67. Charney to Platzman, 22 June 1950 (Charney, 14/451).

68. The Institute for Advanced Study, The Meteorology Group, Progress Report July 1, 1949—June 30, 1950, Contract No. N-6-ori-139 (Charney, 9/304).

69. Thompson to Fletcher, 8 March 1948 (Thompson).

70. Thompson to A. Trakowski, 16 October 1948 (Thompson). Emphasis in the original.

71. Thompson to Paul Worthman, 1 March 1949 (Thompson).

72. Thompson to Trakowski, 16 October 1948 (Thompson).

73. Thompson to Wexler, 24 November 1948 (Thompson).

74. Thompson to Fletcher, 4 June 1948 (Thompson).

75. P. D. Thompson, "Proposed Plan of Air Force Sponsored Research in Meteorology and Closely Allied Sciences,"10 January 1947 (Thompson).

76. Thompson to Fletcher, ca. 1949 (Thompson).

77. P. D. Thompson, "Historical Notes Atmospheric Analysis Laboratory," July 1949 (Thompson).

78. Ibid.

79. Thompson to von Neumann, 11 April 1949 (Thompson).

80. Thompson to R. G. Stone, 25 April 1949 (Thompson).

81. Nebeker, *Calculating the Weather*, 154; P. D. Thompson, "An Interview with Philip Thompson" (conducted by William Aspray, 5 December 1986), Charles Babbage Institute, Minneapolis, MN; AMS TRIP "Interview of Philip D. Thompson, 15–16 December 1987" (conducted by Joseph Tribbia and Akira Kasahara), NCAR Archives, Boulder, CO (on 16).

82. Thompson, "History of Numerical Weather Prediction," and "Maturing of the Science."

83. Thompson to J. Freeman, 20 November 1950 (Thompson).

84. Thompson to Trakowski, 16 August 1948 (Thompson).

85. Thompson, "Notes."

Notes to Chapter 6

1. Reichelderfer to Wexler, 9 August 1950 (Wexler, 5/1950-4).

2. Platzman to Charney, 9 July 1950 (Charney, 14/451).

3. Scorer, "Atmospheric Signal Velocity"; Charney, "Physical Basis."

4. Charney to Rossby, 25 September 1950 (Charney, 14/459).

5. Rossby to Charney, 28 September 1950 (Charney, 14/459).

6. Charney to Rossby, 2 October 1950 (Charney, 14/459).

7. Charney, "Progress in Dynamic Meteorology."

8. Charney to von Neumann, 24 October 1950 (Charney, 15/516).

9. Charney to von Neumann, 21 December 1950 (Charney, 16/517).

10. Wexler and Hermann B. Wobus to Reichelderfer, 24 October 1950 (Wexler, 5/1950).

11. Max A. Eaton to von Neumann, 27 November 1950 with annotations (von Neumann, 15/2). This is the first indication in the archival record that the Air Force wanted to take part in the funding of the Meteorology Project.

12. Rossby to Charney, 7 January 1951 (Charney, 14/459).

13. Wexler to Reichelderfer, 23 January 1951 (Wexler, 5/1951-5).

14. Charney to Wexler, 18 January 1951; Wexler to Charney, 22 January 1951 (Wexler, 5/1951-5).

15. Von Neumann to Oppenheimer, 27 February 1951 (von Neumann, 15/2). In the letter, Platzman is referred to as "H. Platzman" instead of "George Platzman." For more about the role of Holzman in Air Weather Service R&D circles, see Fuller, *Thor's Legions*.

16. Wexler, Namias, G. Brier, J. R. Fulks to Reichelderfer, 21 February 1951 (Wexler, 5/1951-5).

17. Fulks to Reichelderfer, 26 February 1951 (Wexler, 5/1951-2).

18. Fulks to Reichelderfer, 12 March 1951 (Wexler, 5/1951-2).

19. I. R. Tannehill, Memorandum for the Record, 9 July 1951 (Wexler, 5/1951-2).

20. Charney, et al., "Numerical Integration."

21. The Institute for Advanced Study, The Meteorology Group, Progress Report, July 1, 1950 to March 31, 1951, Contract No. N-6-ori-139 (Charney, 9/304). For example, Fjørtoft worked on a simpler version of the three-dimensional quasi-geostrophic equations using an advective assumption that produced results that were superior to the barotropic version. Von Neumann and Charney were exploring relaxation techniques to use with these equations and had successfully tried a modified Liebman method on the two-dimensional model. For a discussion of relaxation techniques, see Haltiner, *Numerical Weather Prediction*.

22. Wexler to L. S. Dederick, 6 April 1951; copy with note from Wexler to von Neumann, same date (Wexler, 5/1951-5).

23. Ray A. Pillivant to Reichelderfer, 11 July 1950 (Wexler, 5/1950); W. Barkley Fritz to von Neumann, 10 April 1951 (von Neumann, 12/3).

24. Von Neumann to Wexler, 9 April 1951 (Wexler, 5/1951-5).

25. Eaton to von Neumann, 10 May 1951 (von Neumann, 12/3).

26. Von Neumann to Eaton, 16 May 1951 (von Neumann, 12/3). Nebeker, *Calculating the Weather*, does not mention Rossby's involvement with the examination of the first ENIAC expedition results or planning for the second ENIAC expedition. This is a crucial omission given Rossby's de-facto leadership of the Meteorology Project from Chicago, Stockholm, or wherever he happened to be.

27. Phillips joined the Project permanently (except for a swap with Bolin when he went to Sweden) in fall 1951. He did extensive work in modeling the general circulation and moved to MIT with Charney at the conclusion of the Project.

28. Charney to Platzman, 12 May 1951 (Charney, 14/451).

29. Platzman to Charney, 17 May 1951 (Charney, 14/451).

30. Von Neumann to Charney, 2 June 1951 (Charney, 16/517).

31. Charney to von Neumann, 13 July 1951 (von Neumann, 15/1).

32. Ibid.

33. Munitalp ('platinum' spelled backwards) was a small foundation that funded basic research in meteorology, primarily areas that its directors thought would not be funded by the military. During a meeting of the board of directors and meteorologists Horace Byers and Sverre Petterssen, in addition to Vincent Schaefer of the General Electric Research Laboratory in Schenectady, New York, foundation leaders decided to place most of their funding with cloud physics (i.e., weather modification) efforts. Before this decision, Munitalp had seriously considered funding international meteorological institutes like the one Rossby was establishing in Stockholm. Proceedings of the First Forum of Munitalp Foundation, Inc., 14 November 1951 (V. J. Schaefer papers, M. E. Grenander Department of Special Collections and Archives, SUNY Albany, B II.3.1U).

34. Charney to von Neumann, 13 July 1951 (von Neumann, 15/1).

35. The Institute for Advanced Study, The Meteorology Group, Progress Report, July 1, 1951 to September 30, 1951, Contract No. N-6-ori-139, Task Order I (Charney, 9/304).

36. Charney to Rossby, 16 November 1951 (Charney, 14/459).

37. Charney to Rossby, 16 November 1951 (Charney, 14/459).

38. Platzman to Phillips and Charney, 20 November 1951 (Charney, 14/451).

39. The Institute for Advanced Study, The Meteorology Group, Progress Report, October 1, 1951 to December 21, 1951, Contract No. N-6-ori-139, Task Order I (Charney, 9/304). Since it turned out to be easier to use potential temperature (θ) instead of the vertical distance as the vertical coordinate, they had first decided to choose θ. However, the ground was not a coordinate surface, so they worked on finding another coordinate that had the advantages of potential temperature and for which the ground was a coordinate surface. They determined the lateral boundary conditions by a heuristic method analogous to the one used in the barotropic model. They took the upper boundary to be a 400K isentropic surface and assumed it was rigid, which allowed them to treat the upper boundary and the earth's surface the same way. After estimating truncation errors for nets of different sizes, they decided to use a horizontal grid spacing of 300 km and a vertical interval equivalent to 150 mb of pressure. The entire grid would cover 4,000 km on a side and contain about 1,200 points. They then

investigated a number of relaxation techniques and settled on a modified Liebman method with an empirically determined constant of overrelaxation.

40. Bolin to Charney, 10 January 1952 (Charney, 4/121).

41. Bolin to Charney, 19 January 1952 (Charney, 4/121).

42. The Institute for Advanced Study, The Meteorology Group, Progress Report, December 22, 1951 to March 31, 1952, Contract No. N-6-ori-139, Task Order I (Charney, 9/304). The team also created a code to convert stream functions, read from the weather charts as decimal numbers, directly into the initial binary vorticities needed by the computer. They reduced the round-off errors by using the absolute vorticity instead of the stream function as the dependent variable, and then solving a Poisson equation to create the stream function from the vorticity field. Their selection of the modified Liebman relaxation method en lieu of Fourier transforms for solving the Poisson equation meant they had to change all the barotropic programs and associated coding, but since they had already chosen the modified Liebman for the three-dimensional model this decision allowed them to try it out on the simpler two-dimensional model first.

43. Rossby to Charney, 5 April 1952 (Charney, 14/459).

44. Charney to Rossby, 9 April 1952 (Charney, 14/459).

45. Rossby to Charney, 16 April 1952 (Charney, 14/459). Hovmöller, a synoptician, was visiting from the Sveriges Metteorologiska Och Hydrologiska Institute (Swedish Meteorological and Hydrographical Institute—SMHI).

46. Wexler to Reichelderfer, 12 May 1952 (Wexler, 5/1952-2). Emphasis in original. Also in attendance for the von Neumann briefing were Joseph Kaplan, Father James Macelwane, S.J., and Ross Gunn of the Geophysics Panel, Science Advisory Board, U.S. Air Force; Bernhard Haurwitz (NYU), Herbert Riehl, and George Platzman (both of Chicago).

47. Reichelderfer to Plans and Program Management Office (P&PMO), 13 June 1952 (Wexler, 5/1951-3).

48. Bolin to Charney, 27 May 1952 (Charney, 4/121).

49. "Violence from the Skies," *New York Times*, 27 November 1950, 23.

50. The Institute for Advanced Study, The Meteorology Project, Summary of Work under contract N-6-ori-139 (1), NR 082-008 during the calendar year 1952 (Charney, 9/304). As an initial simplifying assumption, the team had to consider the motion to be adiabatic because they did not have enough information about non-adiabatic effects. That required the potential temperature to be a conservative quantity that in turn allowed its use as a vertical coordinate in a semi-Lagrangian coordinate system. Fortunately, this led to a simple form of the equation of motion that was well suited for numerical integration.

51. Ibid.

52. P. D. Thompson, request for MATS Transportation, 21 November 1950 (Thompson).

53. Ibid.

54. Thompson to Major C. V. Hendricks, 19 December 1950 (Thompson). The German institutes were located at Göttingen, Mainz, Berlin, Hamburg, and Bad Kissingen.

55. Thompson to Eliassen, 20 December 1950 (Thompson).

56. Thompson, informal notes from European trip, January-February 1951 (Thompson).

57. Thompson to C. N. Touart, 10 May 1951 (Thompson). This shift was not limited to meteorology; see Sapolsky, *Science in the Navy*.

58. Thompson to Chief, Air Weather Service, 27 February 1952 (Thompson).

59. Thompson to Petterssen, 31 March 1952 (Thompson).

60. Thompson to Charney, 31 March 1952 (Thompson).

61. Thompson to Charney, 31 March 1952 (Thompson).

62. Thompson, "An Interview with Philip Thompson" (conducted by William Aspray on 5 December 1986), Charles Babbage Institute, Minneapolis, MN, quoted in Nebeker, *Calculating the Weather*, 154.

63. Thompson to Rossby, 1 April 1952 (Thompson).

64. Ibid.

65. Petterssen to Thompson, 2 April 1952 (Thompson).

66. Rossby to Charney, 5 April 1952 (Charney, 14/459).

67. Touart to Commanding General, Air Research and Development Command, 22 April 1952 (Thompson). Emphasis in original.

68. *GRD Spectrum*, Vol. 1, No. 21, 25 April 1952 (Thompson).

69. Thompson to C. W. Newton, 3 July 1952 (Thompson).

70. Thompson to Charney, 9 July 1952 (Thompson).

71. Record of telephone conversation, Charney and Wexler, 15 July 1952 (Wexler, 5/1952-2).

72. Wexler to Reichelderfer, 16 July 1952 (Wexler, 5/1952-2).

73. Ibid.

74. Charney to Thompson, 16 July 1952 (Thompson).

75. Ibid.

76. Thompson to Bolin, 1 August 1952; Bolin to Thompson, 22 August 1952 (Thompson).

77. Chester Newton to Thompson, 22 July 1952 (Thompson).

78. Wurtele to Thompson, 29 July 1952 (Thompson).

79. Thompson to Wurtele, 4 September 1952 (Thompson).

80. Charney to von Neumann, 17 July 1952 (von Neumann, 15/1).

81. Thompson to Charney, 29 July 1952 (Thompson).

82. Reichelderfer to R. O. Minter (Navy) and Thomas S. Moorman (Air Force), 29 July 1952 (Wexler, 5/1952-1).

83. Aspray, *John von Neumann*, 147. Nebeker's *Calculating the Weather* follows Thompson's "History of Numerical Weather Prediction," which eliminates this entire piece of the story. Likewise, Charney's 1981 interview with George Platzman makes no mention of the trigger for the meeting or the fallout from it. See Lindzen, et al., *The Atmosphere—A Challenge*.

84. Wexler to Reichelderfer, 7 August 1952 (Wexler, 5/1952-2).

85. Minutes of the meeting held at the Institute for Advanced Study, 5 August 1952 concerning practical numerical weather forecasting (von Neumann, 15/4).

86. P. D. Thompson, "Statement to the Conference on Numerical Prediction to be held at Princeton, 5 August 1952" (Charney, 4/135).

87. Wexler to Reichelderfer, 7 August 1952 (Wexler, 5/1952-2).

Notes to Chapter 7

1. "Electronic 'Brain' Planned to Forecast the Weather." A subsequent Science Service article in the 23 August 1952 edition of *Science News Letter* had the modeling topping out at 18,000 feet, still too low. For an insider's recollection of the move from research-oriented to operational NWP and the subsequent efforts of the Joint Numerical Weather Prediction Unit, see Cressman, "Origin."

2. Phillips to Charney, 15 September 1952 (Charney, 14/449).

3. Rex to Phillips, 7 October 1952 (Charney, 14/465).

4. Rex to Charney, 24 November 1952 (Charney, 14/465).

5. Charney to von Neumann, 17 November 1952 (Charney, 16/517).

6. Chief of the Bureau (Reichelderfer) to Assistant Chief (O); SR&F and Scientific Services, 31 October 1952 (Wexler, 5/1952-1).

7. I. R. Tannehill to Reichelderfer, 7 November 1952 (Wexler, 5/1952-1).

8. T. P. Gleiter to Tannehill, 7 November 1952 (Wexler, 5/1952-1). This is an important and heretofore unexplored instance of national styles in science.

9. Von Neumann to A. Ångström, 14 November 1952 (Charney, 16/517). Ångström spent much of his career studying radiation. In the 1920s, he worked with the Smithsonian's Abbot measuring the solar constant. Ångström came from a long line of scientists. His grandfather, Anders Jonas Ångström, lent the family name to a non-SI unit of wavelength: 1 ångström (Å) = 10^{-10} meters. SI units (Système International d'unités) are those based on the gram/centimeter/second or kilogram/meter/second units of measurement.

10. Von Neumann to Ångstrom, 2 December 1952 (Charney, 16/517).

11. Rossby to Charney, 12 January 1953 (Charney, 14/460).

12. Von Neumann to Ångström, 3 February 1953 (Charney, 16/517).

13. Bolin to Charney, 29 January 1953 (Charney, 4/121).

14. Charney to Bolin, 3 February 1953 (Charney, 4/121).

15. Bolin to Charney, 14 February 1953 (Charney, 4/121).

16. Charney to Bolin, 17 February 1953 (Charney, 4/121).

17. The Institute for Advanced Study, The Meteorology Project, Quarterly Progress Report, January 1, 1953 to March 31, 1953, Contract No. N-6-ori-139 (1), NR 082-008 (Charney, 9/305).

18. The Institute for Advanced Study, The Meteorology Project, Quarterly Progress Report, April 1, 1953 to June 31, 1953, Contract No. N-6-ori-139 (1), NR 082-008 (Charney, 9/305).

19. Ibid.

20. Ibid.

21. The Institute for Advanced Study, Meteorology Project, Progress Report 1 July 1953 to 31 March 1954, Contract No. N-6-ori-139(1), NR 082-008 (Charney, 9/305).

22. Ibid.

23. Chief, Weather Bureau (Reichelderfer) to Joseph Smagorinsky, 8 September 1952 (Wexler, 32/NWP).

24. Joseph Smagorinsky to R. N. Culnan, 30 September 1952 (Wexler, 32/NWP).

25. Joseph Smagorinsky to Wexler, 5 March 1953 (Wexler, 32/NWP).

26. Chief, Weather Bureau (Reichelderfer) to Assistant Chief, Scientific Services (Wexler) and P&PMO (Tannehill), 24 March 1953 (Wexler, 32/NWP). Costs of maintaining an upper-air station included buying supplies (large, specially made balloons,

hydrogen or helium gas to inflate them, and the instrument packages [radiosondes] they carried) to sustain two launches per day, maintaining the radio tracking and communications circuits, and employing specially trained observers. In some places, the cost per launch (in 1953 dollars) was $100—a considerable expense.

27. Joseph Smagorinsky to Chief, Weather Bureau (Reichelderfer) via Wexler, 29 May 1953 (Wexler, 32/NWP).

28. Ibid.

29. Proposal for a Study Project on Automatic Procurement and Processing of Data (APPOD), Weather Bureau, 15 July 1953 (Wexler, 32/NWP).

30. Ibid.

31. Ibid.

32. Rex to Phillips, 7 October 1952 (Charney, 14/465); Rex to Charney, 24 November 1952 (Charney, 14/465).

33. Joseph Smagorinsky to Chief, Weather Bureau (Reichelderfer) via Wexler, 29 May 1953 (Wexler, 32/NWP).

34. Supporting Paper by US Weather Bureau Member (Reichelderfer) on Numerical Weather Prediction for the Joint Meteorological Committee, 10 June 1953 (Wexler, 32/ NWP). Fuller, *Thor's Legions*, claims the Air Force was pushing for a joint numerical weather prediction unit, that Reichelderfer and Wexler did not think NWP was ready to go operational, and that the position paper presented by Reichelderfer to the JMC was actually written by the Air Weather Service in spring 1953. The archival evidence shows otherwise. The Weather Bureau had already started to move into operational NWP within a month of the August 1952 meeting—a meeting at which the Air Force *declined* to go operational with a joint unit. Indeed, the AWS did make overtures to other participants in the spring of 1953, but only because it feared being left out of the program. See Fuller, *Thor's Legions*, 222 (and footnote 35).

35. Minutes of the sixth meeting of the Ad Hoc Committee on Numerical Weather Prediction held 11 August 1953. Members of the committee: Chair: Commander Daniel F. Rex, USN (OPNAV); Majors W. H. Best and T. H. Lewis, USAF, (Air Weather Service); Drs. H. Wexler and J. Smagorinsky (US Weather Bureau). Based on Rex's letter to von Neumann, the committee invited von Neumann as well as meteorologists Charney, Gilchrist, Berggren, and computer engineer Bigelow from IAS to provide advice. Rex to von Neumann, 22 July 1953 (von Neumann, 15/4).

36. Report by the Ad Hoc Committee on Numerical Weather Prediction to the Joint Meteorological Committee on Joint Numerical Weather Prediction (NWP-10-53), 12 August 1953 (Wexler, 6/1953). The technical consultants were from (1) IAS: John von Neumann, Roy Berggren, Julian Bigelow, Jule Charney, and Bruce Gilchrist; (2) IBM: C. C. Hurd, George W. Petrie, and John Sheldon; (3) University of Chicago: Sverre

Petterssen and George W. Platzman; (4) Bureau of Standards, Charles B. Tompkins. Major P. D. Thompson, USAF, was an "unaffiliated member."

37. Ibid.

38. Enclosure: Plan for Joint Numerical Weather Prediction Unit, 12 August 1953 (Wexler, 6/1953).

39. Ibid.

40. Ibid.

41. Ibid., Appendix, p. 9, paragraph 1b-d, 2, and 3.

42. Ibid., Appendix, section III, p. 1, paragraph 1.

43. Ceruzzi, *History of Modern Computing*, 34.

44. Ibid.

45. Enclosure: Plan for Joint Numerical Weather Prediction Unit, 12 August 1953 (Wexler, 6/1953).

46. Ibid., Appendix, section IV, Para 1 and 3.

47. Reichelderfer to Division Heads, 21 August 1953 (Wexler, 32/NWP).

48. Wexler to Reichelderfer, 27 August 1953 (Wexler, 6/1953).

49. Wexler to Reichelderfer, 3 September 1953 (Wexler, 6/1953).

50. Reichelderfer to Wexler, 8 September 1953 (Wexler, 6/1953).

51. JCS/JMC Ad Hoc Group for Establishment of a Joint Numerical Weather Prediction Unit (JNWP-11-53) Minutes of the 1st meeting held 22 September 1953 (Wexler, 6/1953).

52. JCS/JMC Ad Hoc Group for Establishment of a JNWPU (JNWP-12-53) Minutes of the 3rd meeting held 12 October 1953 (Wexler, 6/1953). Just how Remington Rand expected to run a competitive test without a computer is a bit of a mystery. According to historian of computing Paul Ceruzzi, Remington Rand, which had acquired both UNIVAC and Engineering Research Associates (ERA), "did not fully understand what it had bought." Consequently, it did not know how to market its computers. See Ceruzzi, *History of Modern Computing*, 45.

53. JCS/JMC Ad Hoc Group for Establishment of a Joint Numerical Weather Prediction Unit (JNWP-11-53) Minutes of the second meeting held 6 October 1953 (Wexler, 6/1953).

54. JCS/JMC Ad Hoc Group for Establishment of a Joint Numerical Weather Prediction Unit (JNWP-11-53) Minutes of the third meeting held 12 October 1953 (Wexler, 6/1953).

55. JCS/JMC Ad Hoc Group for Establishment of a JNWPU: Minutes of the 4th meeting held 28 October 1953 (JNWP-18-53) (Wexler, 6/1953).

56. Joseph Smagorinsky to H. Wexler, 19 August 1953 (Wexler, 32/NWP).

57. Tentative Report to the Ad Hoc Group for Establishment of a Joint Numerical Weather Prediction Unit, "The ERA 1103 and the IBM 701," 26 January 1954 (Charney, 16/522). Smagorinsky and Goldstine wrote this unsigned report.

58. Minutes of the First Meeting of the Technical Advisory Group, Ad Hoc Group for Establishment of a Joint Numerical Weather Prediction Unit (JNWP-6-54) (Wexler, 6/1953). The list of technical advisors had grown and at this point included von Neumann, Charney and Bigelow (all from IAS), Charles B. Tompkins (Bureau of Standards), Joseph J. Eachus (National Security Agency), mathematician Mina Rees (Hunter College), and C.V. L. Smith (ONR).

59. Reichelderfer to Smagorinsky, 19 February 1954 (Wexler, 6/1954).

60. JCS/JMC Ad Hoc Group for Establishment of a JNWPU: Minutes of the 11th meeting held on 11 March 1954 (JNWP-16-54) (Wexler, 32/NWP).

61. JCS/JMC Ad Hoc Group for Establishment of a JNWPU: Minutes of the 14th Meeting held on 28 May 1954 (JNWP-19-54) (Wexler, 32/NWP).

62. Final Report of the Ad Hoc Group for Establishment of a Joint Numerical Weather Prediction Unit dated 30 June 1954 (Enclosure [1] to the Minutes of the 15th meeting held 1 July 1954) (Wexler, 32/NWP).

63. D. F. Rex, Chairman, Ad Hoc Group Secretary, Joint Meteorological Committee (JNWP-2-53), 22 September 1953 (Wexler, 6/1953). Cressman would later become the director of the National Weather Service (1965–1979).

64. Report by the Ad Hoc Committee on Numerical Weather Prediction to the Joint Meteorological Committee on Joint Numerical Weather Prediction, Enclosure: Plan for Joint Numerical Weather Prediction Unit, Appendix, section III, pp. 12–13, 12 August 1953 (Wexler, 6/1953).

65. Final Report of the Ad Hoc Group for Establishment of a Joint Numerical Weather Prediction Unit, 30 June 1954 (Enclosure [1] to the Minutes of the 15th meeting held 1 July 1954); JCS/JMC Ad Hoc Group for Establishment of a JNWPU: Minutes of the 6th Meeting held on 7 December 1953 (JNWP-27-53); final version JCS/JMC Ad Hoc Group for Establishment of a JNWPU: Minutes of the 8th Meeting held on 28 January 1954 (JNWP-7-54) (Wexler, 32/NWP).

66. Final Report of the Ad Hoc Group for Establishment of a Joint Numerical Weather Prediction Unit dated 30 June 1954 (Enclosure [1] to the Minutes of the 15th meeting held 1 July 1954) (Wexler, 32/NWP).

67. JCS/JMC Ad Hoc Group for Establishment of a JNWPU, Minutes of the 4th Meeting held 28 October 1953 (JNWP-18-53) (Wexler, 6/1953). The JMC formally designated the Weather Bureau as the administrative authority on 17 February 1954.

68. Draft JCS/JMC Ad Hoc Group for Establishment of JNWPU, Minutes of the 8th Meeting held on 28 January 1954 (Wexler, 32/NWP).

69. Final Report of the Ad Hoc Group for Establishment of a Joint Numerical Weather Prediction Unit, 30 June 1954 (Enclosure [1] to the Minutes of the 15th meeting held 1 July 1954) (Wexler, 32/NWP). The estimated personnel expenses obtained by using mid-grade Civil Service salaries came to $191,525. The IBM 701 (and peripherals) lease would be approximately $97,000 for the four months from March through June 1955. Additional cost account items included travel, phones, utilities, printing, and office supplies for a total of approximately $22,000.

70. Phillips to Charney, undated, ca. fall 1953 (Charney, 14/449).

71. Thompson to Chief, Atmospheric Analysis Laboratory, ca. November 1953 (Thompson).

72. Phillips to Charney, 19 December 1953 (Charney, 14/449).

73. JCS/JMC Ad Hoc Group for Establishment of a JNWPU: Minutes of the 11th Meeting held on 11 March 1954 (JNWP-16-54) (Wexler, 32/NWP).

74. Rossby to Charney, 16 June 1954 (Charney, 14/460). For details of numerical weather prediction outside the United States, particularly Sweden, see Persson, "NWP in Sweden," and "Twenty countries."

75. "Numerical Weather Prediction," address on WGY Science Forum by Joseph Smagorinsky, US Weather Bureau, Washington, DC, 20 May 1953 (Wexler, 32/NWP).

76. Ibid.

77. Smagorinsky to Reichelderfer via Wexler, 29 May 1953 (Wexler, 32/NWP). *Fortune's* article "Tomorrow's Weather" pitted the Weather Bureau's extended forecasting section against the private forecasting firm of "war forecaster" Irving P. Krick, whom Reichelderfer detested. It also addressed the weather modification work of both Krick and Langmuir.

78. Emphasis added. Article ("Weather by Giant 'Brain'") by Ann Ewing, Science Service Staff Writer on NWP, October 1953 (Wexler, 32/NWP).

79. Office of Naval Research Press Release "Electronic Weather Forecasting," 23 November 1953 (von Neumann, 15/2).

80. JCS/JMC Ad Hoc Group for Establishment of a JNWPU: Minutes of the 6th Meeting held on 7 December 1953 (JNWP-27-53) (Wexler, 32/NWP).

81. Wexler to Reichelderfer, 9 December 1953 (Wexler, 32/NWP).

82. JCS/JMC Ad Hoc Group for Establishment of a JNWPU: Minutes of the 9th Meeting held on 5 February 1954 (JNWP-12-54) (Wexler, 32/NWP).

83. Von Neumann to Mr. Carol Spica, 5 March 1954 (Charney, 16/517).

84. Activities of the Joint Numerical Weather Prediction Unit, 1 July to 1 October 1954 (Wexler, 32/NWP).

85. The original members of the JMC/NWP were Captain W. E Oberholtzer, Jr., USN, Lt. Colonel H. H. Bedke, USAF, and Dr. H. Wexler, US Weather Bureau.

86. Terms of Reference, Joint Meteorological Committee, Ad Hoc Committee on Numerical Weather Prediction (JMC/NWP), Enclosure to Memorandum for the Members, Joint Meteorological Committee (JMC-130-54) of 10 November 1954 (USWB, JCS/JMC).

87. JMC/NWP Minutes of 1st Meeting held 29 November 1954 (USWB, JCS/JMC).

88. JMC/NWP Minutes of 2nd Meeting of 13 December 1954 (USWB, JCS/JMC).

89. JMC/NWP Minutes of 1st Meeting held 29 November 1954 (Wexler, 32/NWP). The forerunner of the Geophysical Fluid Dynamics Laboratory (GFDL) was not proposed until October 1955, so it is unlikely that Cressman was assuming that another Weather Bureau-related organization would take over the programming duties.

90. Ibid.

91. JMC 60/15.5; Joint Meteorological Committee, Coordination Between the Joint Numerical Weather Prediction Unit and the National Weather Analysis Center, 16 December 1954 (USWB, JCS/JMC).

92. Minutes of the Joint Meteorological Committee 340th Meeting held 21 December 1954 (USWB, JCS/JMC).

93. Ibid.

94. Minutes of the Joint Meteorological Committee 341st Meeting held 18 January 1955 (USWB, JCS/JMC).

95. Minutes of the 344th meeting of the JMC held 3 May 1955 (USWB, JCS/JMC).

96. "The Electronic Brain."

97. "Suggested Remarks for Mr. Little at the Joint Numerical Weather Prediction Unit Opening Ceremony May 6, 1955," 4 May 1955 (Reichelderfer, 2/10).

Notes to Chapter 8

1. For short discussions that focus on model development within the JNWPU and after written by participants, see Cressman, "Origin"; Shuman "History of Numerical Weather Prediction"; and Thompson, "History of Numerical Weather Prediction," and "Maturing of the Science." See also Nebeker, *Calculating the Weather*, 160–161.

2. Mr. Vernon's statement, Minutes of the SC/NAWAC's 19th meeting held 26 March 1957 (USWB, JCS/JMC).

3. Minutes of the 5th meeting JMC/NWP held 28 September 1955; minutes of the 6th meeting JMC/NWP held 27 December 1955 (USWB, JCS/JMC).

4. Ibid.

5. Comments on the JMC's 352nd meeting held 20 March 1956 (USWB, JCS/JMC).

6. Minutes of the JMC's 355th meeting held 18 June 1956 (USWB, JCS/JMC).

7. Cressman to J. Eberly, 21 October 1948; minutes of and Weather Bureau comments on the JMG's 371st meeting held 13 January 1959; minutes of and Weather Bureau comments on the JMG's 372nd meeting held 24 February 1959 (USWB, JCS/JMG).

8. Minutes of the JMG's 373rd meeting held 7 April 1959 (USWB, JCS/JMG).

9. Minutes of the JMG's 381st meeting held 16 February 1960 (USWB, JCS/JMG).

10. Minutes of the 147th Panel/WP meeting held 17 January 1961 (USWB, JCS/JMG).

11. Enclosure to Item 2, Briefing by the Director, NMC to the JMG on Numerical Weather Unit Matters dated 8 August 1961 (USWB, JCS/JMG).

12. The author is a retired Navy meteorologist whose first assignment as a recently graduated mathematics major was as the assistant computer officer at the US Fleet Weather Central, Rota, Spain, from January 1974 through June 1976. She draws on her 21 years of experience in meteorological and oceanographic support to military customers and coordination with the National Weather Service in writing this section. On practice in scientific communities see for instance, Kohler, *Lords of the Fly*.

13. For a discussion of weather and climate modeling throughout the world, see J. Houghton, "Predictability." For brief historical discussions of early work in Sweden, see Wiin-Nielsen, "Numerical Weather Prediction," and Persson, "NWP in Sweden"; in Germany, see Reiser, "Development"; on over twenty different countries, see Persson, "Twenty countries." For a brief history of the ECMWF, see Bengtsson, "Medium-Range Forecast." For a longer history, see Woods, *Medium-Range Weather Prediction*.

14. Phillips, "General Circulation."

15. Quiñones, "Pioneering Meteorologist."

16. Taba, *Bulletin Interviews*, 154–155.

17. Smagorinsky, "General Circulation."

18. Taba, *Bulletin Interviews*, 156.

19. Ibid.

20. See R. J. Fleming et al., "Global Weather Experiment."

21. Harper et al., "Symposium."

22. Ibid.

23. See Weir, *Ocean in Common*.

24. See DeVorkin, *Vengeance*.

25. Von Neumann to Ångstrom, 14 November 1952 (Charney, 16/517).

26. See Dupree, *Science in the Federal Government*, chapter 9.

27. See Oreskes and Doel, "Geophysics."

28. See, e.g., Oreskes et al., "Verification, validation, and confirmation"; Oreskes and Belitz, "Philosophical Issues"; Kwa, "Modelling Technologies"; and Hacker, "Nuclear Weapons."

Bibliography

Manuscript Sources

Library of Congress, Manuscript Division, Washington (Edward L. Bowles Papers, Francis Wilton Reichelderfer Papers, Harry Wexler Papers, John von Neumann Papers)

Institute Archives and Special Collections, MIT Libraries, Cambridge, Massachusetts (Jule Gregory Charney Papers, MC 184; University Meteorological Committee Papers, MC 511)

National Academy of Sciences Archives, Washington (Executive Board, Science Advisory Board, NAS Organizations 1938–1939, Agencies and Departments)

California Institute of Technology Archives, Pasadena (Robert A. Millikan Papers)

M. E. Grenander Department of Special Collections and Archives, State University of New York, Albany (Vincent J. Schaefer Papers)

National Center for Atmospheric Research Archives, Boulder (Philip D. Thompson Papers)

National Archives and Records Administration II, College Park, Maryland (Weather Bureau Papers, RG 27)

Manuscripts, Special Collections, and University Archives, University of Washington, Seattle (WU President, Accession 71–34)

Books and Articles

Abbe, Cleveland. "The Progress of Science as Illustrated by the Development of Meteorology." In Annual Report of Smithsonian Institution, 1907.

Abbot, Charles G. "The Weather and Radiation." *Yale Review* 16 (1927): 485–500.

Appel, Toby. *Shaping Biology: The National Science Foundation and American Biological Research, 1945–1975*. Johns Hopkins University Press, 2000.

Ashford, Oliver M. *Prophet—or Professor? The Life and Work of Lewis Fry Richardson*. Hilger, 1985.

Aspray, William. *John von Neumann and the Origins of Modern Computing*. MIT Press, 1990.

Auerbach, Lewis E. "Scientists in the New Deal: A Pre-War Episode in the Relations between Science and Government in the United States." *Minerva* 3 (1965): 457–482.

Barton, Bruce. "What the Weather Does to You." *American Magazine* 97 (June 1924): 38–39.

Bates, Charles C. "The Formative Rossby-Reichelderfer Period in American Meteorology 1926–1940." *Weather and Forecasting* 4 (1989): 593–603.

Bates, Charles C., and John F. Fuller. *America's Weather Warriors: 1814–1985*. Texas A&M University Press, 1986.

Battan, Louis J. *Harvesting the Clouds: Advances in Weather Modification*. Doubleday, 1969.

Baxter, James Phinney. *Scientists against Time*. Little, Brown, 1948.

Beals, E. A. "Is It Possible to Predict California's Rainfall Several Months in Advance?" *Bulletin of the American Meteorological Society* 18 (1927): 103–107.

Bean, Louis H. "Weather and Crop Research under Bankhead-Jones Fund: Progress Report." *Bulletin of the American Meteorological Society* 17 (1936): 288–292.

Bengtsson, Lennart. "The Development of Medium Range Forecasts." In *50th Anniversary of Numerical Weather Prediction Commemorative Symposium, 9–10 March 2000, Book of Lectures*, ed. A. Spekat. Deutsche Meterologische Gesellschaft, 2000.

Bergeron, Tor. "The Young Carl-Gustaf Rossby." In *The Atmosphere and the Sea in Motion: Scientific Contributions to the Rossby Memorial Volume*, ed. B. Bolin. Rockefeller Institute Press, 1959.

Berkeley, E. C. *Giant Brains, or Machines That Think*. Wiley, 1949.

Billings, John, Jr. "Is the Sun Fickle?" *Independent* 117 (1926): 269–271.

Blochman, L. E. "The Difficulties of Long-Range Forecasting." *Bulletin of the American Meteorological Society* 10 (1929): 222–223.

Boesen, Victor. *Storm: Irving Krick vs. the U.S. Weather Bureaucracy*. Putnam, 1978.

Bolin, Bert. "Carl-Gustaf Rossby in Memoriam." *Tellus* (1957): 257–258.

Bowie, Edward H. "Presidential Address." *Bulletin of the American Meteorological Society* 24 (1943): 34–38.

Brooks, Charles F. "Collegiate Instruction in Meteorology." *Monthly Weather Review* 46 (1918): 555–560.

Brooks, Charles F. "General Extent of Collegiate Instruction in Meteorology and Climatology in the United States." *Monthly Weather Review* 47 (1919): 169–170.

Brooks, Charles F. "From the Committees: Meteorological Instruction." *Bulletin of the American Meteorological Society* 1 (1920): 15–16.

Brooks, Charles F. "Meteorological Instruction." *Bulletin of the American Meteorological Society* 1 (1920): 56–57.

Brooks, Charles F. "Reclassification." *Bulletin of the American Meteorological Society* 3 (1922): 163–164.

Brooks, Charles F. "An Outline of the Study of World Weather and Long Range Weather Forecasting." *Bulletin of the American Meteorological Society* 8 (1927): 31–32.

Brooks, Charles F. "Our Society's First Decade." *Bulletin of the American Meteorological Society* 11 (1930): 8–12.

Brooks, Charles F. "Many Government Meteorologists Retired." *Bulletin of the American Meteorological Society* 13 (1932): 153–154.

Brooks, Charles F. "*Monthly Weather Review* Halved." *Bulletin of the American Meteorological Society* 13 (1932): 154.

Brophy, Leo P., and George J. B. Fisher. *United States Army in World War II: The Technical Services: The Chemical Warfare Service: Organizing for War*. Department of the Army, 1959.

Brophy, Leo P., Wyndham D. Miles, and Rexmond C. Cochrane. *United States Army in World War II: The Technical Services: The Chemical Warfare Service: From Laboratory to Field*. Department of the Army, 1959.

Burchard, John. *Q.E.D.: M.I.T. in World War II*. Technology Press, 1948.

Byers, Horace R. "Carl-Gustaf Arvid Rossby." *Bulletin of the American Meteorological Society* 39 (1958): 98–99.

Byers, Horace R. "Recollections of the War Years." *Bulletin of the American Meteorological Society* 51 (1970): 214–217.

Byers, Horace R. "Carl-Gustaf Rossby, the Organizer." In *The Atmosphere and the Sea in Motion: Scientific Contributions to the Rossby Memorial Volume*, ed. B. Bolin. Rockefeller Institute Press, 1959.

Byers, Horace R. "Carl-Gustaf Arvid Rossby." In *Biographical Memoirs*, volume 34. National Academy of Sciences, 1960.

Byers, Horace R. "The Founding of the Institute of Meteorology at the University of Chicago." *Bulletin of the American Meteorological Society* 57 (1976): 1343–1345.

Byers, Horace R., and L. F. Page. "Conference on Long-Range Forecasting Held at the Department of Agriculture, Washington, D.C., April 30, 1937." *Bulletin of the American Meteorological Society* 18 (1937): 371–373.

Byers, Horace R., J. Kaplan, and E. J. Minser. "The Teaching of Meteorology in Colleges and Universities; Recommendations of the Committee on Meteorological Education of the American Meteorological Society." *Bulletin of the American Meteorological Society* 27 (1946): 95–98.

Campbell-Kelly, Martin, and William Aspray. *Computer: A History of the Information Machine*. Basic Books, 1996.

Cartwright, Gordon D. and Charles H. Sprinkle. "A History of Aeronautical Meteorology: Personal Perspectives, 1903–1995." In *Historical Essays on Meteorology 1919–1995: The Diamond Anniversary History Volume of the American Meteorological Society*, ed. J. Fleming. American Meteorological Society, 1996.

Ceruzzi, Paul E. *A History of Modern Computing*. MIT Press, 1998.

Chapman, Sydney. Introduction to *Weather Prediction by Numerical Process*, by Lewis Fry Richardson. Dover, 1965.

Charney, Jule. "The Dynamics of Long Waves in a Baroclinic Westerly Current." *Journal of Meteorology* 4 (1947): 135–162.

Charney, Jule. "On a Physical Basis for Numerical Prediction of Large-Scale Motions in the Atmosphere." *Journal of Meteorology* 6 (1949): 371–385.

Charney, Jule. "Progress in Dynamic Meteorology." *Bulletin of the American Meteorological Society* 31 (1950): 231–236.

Charney, Jule, and Arnt Eliassen. "A Numerical Method for Predicting Perturbations of the Middle Latitude Westerlies." *Tellus* 2 (1949): 38–54.

Charney, Jule, R. Fjörtoft, and J. von Neumann. "Numerical Integration of the Barotropic Vorticity Equation." *Tellus* 3 (1951): 248–257.

Charnock, Henry. "Ocean-Atmosphere Interactions." In *Sciences of the Earth: An Encyclopedia of Events, People, and Phenomena*, ed. G. Good. Garland, 1998.

Compton, Karl T. Report of the Science Advisory Board, July 31, 1933 to September 1, 1934, 20 September 1934.

Conner, Patrick. "Sunshine for Saturday, *Please!*" *American Magazine* 106 (July 1928): 105.

Conover, John H. *The Blue Hill Meteorological Observatory: The First 100 Years*. American Meteorological Society, 1990.

Conover, Milton. *The Office of Experiment Stations: Its History, Activities and Organization*. Johns Hopkins University, 1924.

Cotton, William R. and Roger A. Pielke. *Human Impacts on Weather and Climate*. Cambridge University Press, 1995.

Cox, Henry J. "Curious Ways in Which the Weather Affects Business." *American Magazine* 94 (August 1922): 54–55, 150.

Cressman, George P. "The Origin and Rise of Numerical Weather Prediction." In *Historical Essays on Meteorology, 1919–1995: The Diamond Anniversary History Volume of the American Meteorological Society*, ed. J. Fleming. American Meteorological Society, 1996.

Culver, John C. and John Hyde. *American Dreamer: A Life of Henry A. Wallace*. Norton, 2000.

Dennis, Michael A. "Historiography of Science: An American Perspective." In *Science in the Twentieth Century*, ed. J. Krige and D. Pestre. Harwood, 1997.

DeVorkin, David H. "Organizing for Space Research: The V-2 Rocket Panel." *Historical Studies in the Physical and Biological Sciences* 18 (1987): 1–24.

DeVorkin, David H. "Defending a Dream." *Journal for the History of Astronomy* 21 (Part 1) (1990): 121–136.

DeVorkin, David H. *Science with a Vengeance: How the Military Created the U.S. Space Sciences after World War II*. Springer, 1992.

DeVorkin, David H. "Charles Greely Abbot." In *Biographical Memoirs*, volume 73. National Academy of Sciences, 1998.

DeVorkin, David H. *The American Astronomical Society's First Century*. American Institute of Physics, 1999.

DeVorkin, David H. *Henry Norris Russell: Dean of American Astronomers*. Princeton University Press, 2000.

Dieudonne, J. "John von Neumann." In *Dictionary of Scientific Biography*, volume 14, ed. C. Gillispie. Scribner, 1990.

Doel, Ronald E. *Solar System Astronomy in America: Communities, Patronage, and Interdisciplinary Science, 1920–1960*. Cambridge University Press, 1996.

Doel, Ronald E. "The Earth Sciences and Geophysics." In *Science in the Twentieth Century*, ed. J. Krige and D. Pestre. Harwood, 1997.

Doel, Ronald E. "Geophysics in Universities." In *Sciences of the Earth: An Encyclopedia of Events, People, and Phenomena*, ed. G. Good. Garland, 1998.

Dupree, A. Hunter. "The Great Instauration of 1940: The Organization of Scientific Ideas for War." In *The Twentieth Century Sciences: Studies in the Biography of Ideas*, ed. G. Holton. Norton, 1972.

Dupree, A. Hunter. *Science in the Federal Government: A History of Policies and Activities*. Johns Hopkins University Press, 1986.

Edwards, Paul N. "Representing the Global Atmosphere: Computer Models, Data, and Knowledge about Climate Change." In *Changing the Atmosphere: Expert Knowledge and Environmental Governance*, ed. C. Miller and P. Edwards. MIT Press, 2001.

E. E. F., "The Weather and Our Feelings." *Forum* 74 (September 1925): 390–392.

Elliot, Clark A. and Margaret W. Rossiter, eds. *Science at Harvard University: Historical Perspectives*. Lehigh University Press, 1992.

Farber, Paul Lawrence. *Finding Order in Nature: The Naturalist Tradition from Linnaeus to E. O. Wilson*. Johns Hopkins University Press, 2000.

Fassig, Oliver L. "A Signal Corps School of Meteorology." *Monthly Weather Review* 46 (1918): 560–561.

Fassig, Oliver L. "The Weather Bureau." *Bulletin of the American Meteorological Society* 14 (1933): 111–112.

Feldman, Theodore S. "Late Enlightenment Meteorology." In *The Quantifying Spirit in the Eighteenth Century*, ed. T. Frängsmyr et al. University of California Press, 1990.

Feldman, Theodore S. "Meteorology, Medical." In *Sciences of the Earth: An Encyclopedia of Events, People, and Phenomena*, ed. G. Good. Garland, 1998.

Flammer, Philip Maynard. Meteorology in the United States Army, 1917–1935. Master's thesis, George Washington University, 1958.

Fleagle, Robert G. *Weather Modification: Science and Public Policy*. University of Washington Press, 1969.

Fleming, James Rodger. *Meteorology in America, 1800–1870*. Johns Hopkins University Press, 1990.

Fleming, James Rodger. *Historical Perspectives on Climate Change*. Oxford University Press, 1998.

Fleming, R. J., T. J. Kaneshige, and W. E. McGovern. "The Global Weather Experiment: 1. The Observational Phase Through the First Observing Period." *Bulletin of the American Meteorological Society* 60 (1979): 649–659.

Forman, Paul. "Behind Quantum Electronics: National Security as a Basis for Physical Research in the United States, 1940–1960." *Historical Studies in the Physical and Biological Sciences* 18 (1987): 149–229.

Friedman, Robert Marc. *Appropriating the Weather: Vilhelm Bjerknes and the Construction of Modern Meteorology*. Cornell University Press, 1989.

Frisinger, H. Howard. *The History of Meteorology: to 1800*. American Meteorological Society, 1983.

Fuller, John F. *Thor's Legions: Weather Support to the U.S. Air Force and Army, 1937–1987.* American Meteorological Society, 1990.

Geison, Gerald L. "Scientific Change, Emerging Specialties, and Research Schools." *History of Science* 19 (1981): 20–40.

Geison, Gerald L., and Frederic L. Holmes, eds. *Research Schools: Historical Reappraisals. Osiris*, second series, 8 (1995).

Genuth, Joel. "Groping towards Science Policy in the United States in the 1930s." *Minerva* 35 (1987): 238–268.

Geschwind, Carl-Henry. *California Earthquakes: Science, Risk, and the Politics of Hazard Mitigation.* Johns Hopkins University Press, 2001.

Gingerich, Owen. *The General History of Astronomy*, volume 4A: *Astrophysics and Twentieth-Century Astronomy to 1950.* Cambridge University Press, 1984.

Glickman, Todd S., ed. *Glossary of Meteorology,* second edition. American Meteorological Society, 2000.

Goldstine, H. H., and A. Goldstine. "The Electronic Numerical Integrator and Computer (ENIAC)." *Mathematical Tables and Other Aids to Computation* 2 (1946): 97–110.

Good, Gregory A. "The Rockefeller Foundation, the Leipzig Geophysical Institute, and National Socialism in the 1930s." *Historical Studies in the Physical and Biological Sciences* 21 (1991): 299–316.

Good, Gregory A., ed. *The Earth, the Heavens and the Carnegie Institution of Washington.* American Geophysical Union, 1994.

Goodstein, Judith R. *Millikan's School: A History of the California Institute of Technology.* Norton, 1991.

Green, Fitzhugh. *A Change in the Weather.* Norton, 1977.

Gregg, Willis Ray. "History of the Application of Meteorology to Aeronautics with Special Reference to the United States." *Monthly Weather Review* 61 (1933): 165–169.

Gregg, Willis Ray. "Introductory Remarks." *Bulletin of the American Meteorological Society* 20 (1939): 129–132.

Hacker, Barton. "Nuclear Weapons." In *The Oxford Companion to United States History*, ed. P. Boyer et al. Oxford University Press, 2001.

Hagen, Joel B. "Clementsian Ecologists: The Internal Dynamics of a Research School." *Osiris*, second series, 8 (1993): 178–195.

Halacy, D. S., Jr. *The Weather Changers.* Harper & Row, 1968.

Hale, P. G. "The Navy's Part in Modern Aerological Developments in the United States." *Bulletin of the American Meteorological Society* 16 (1935): 114–116.

Hallion, Richard P. *Legacy of Flight: The Guggenheim Contribution to American Aviation.* University of Washington Press, 1977.

Haltiner, G. J. *Numerical Weather Prediction.* Wiley, 1971.

Harding, G. E. "Meteorology and Climatology at a Teachers College." *Bulletin of the American Meteorological Society* 16 (1935): 40–42.

Haurwitz, Bernhard. "Meteorology in the 20th Century: A Participant's View," Parts 1–4. *Bulletin of the American Meteorological Society* 66 (1985): 282–291, 424–431, 498–504, 628–633.

Hayes, Brian. "The Weatherman." *American Scientist* 89 (2001): 10–14.

Henson, Pamela M. "The Comstock Research School in Evolutionary Entomology." *Osiris*, second series, 8 (1993): 159–177.

Hess, Wilmot N. *Weather and Climate Modification.* Wiley, 1974.

Hirt, Paul W. *A Conspiracy of Optimism: Management of the National Forests Since World War Two.* University of Nebraska Press, 1994.

Houghton, H. G. "The President's Page." *Bulletin of the American Meteorological Society* 27 (1946): 191–192.

Houghton, John. "The Bakerian Lecture, 1991: The Predictability of Weather and Climate." *Philosophical Transactions: Physical Sciences and Engineering* 337 (1991): 521–572.

Hughes, Patrick. *A Century of the Weather Service: A History of the Birth and Growth of the National Weather Service, 1870–1970.* Gordon and Breach, 1970.

Hughes, Thomas P. *American Genesis: A Century of Invention and Technological Enthusiasm, 1870–1970.* Viking, 1989.

Humphreys, W. J. "The Claims of Meteorology for a Place in College." *Bulletin of the American Meteorological Society* 10 (1929): 164–166.

Huntington, Ellsworth. "From the Committees: Physiological Meteorology." *Bulletin of the American Meteorological Society* 1 (1920): 18–19.

Huntington, Ellsworth. "The New Astrology." *Century* 110 (May 1925): 106–114.

Huntington, Ellsworth. "What the Weather Does to Us." *Scribners* 79 (June 1926): 571–577.

Jacobs, Woodrow C. A Survey of Instruction in Meteorology in the Colleges of the United States. Master's thesis, University of Southern California, 1934.

Jones, E. Lester. Annual Report of the Director, U.S. Coast and Geodetic Survey, to the Secretary of Commerce for the Fiscal Year ended June 30, 1922. Government Printing Office, 1922.

Kargon, Robert. "The Conservative Mode: Robert A. Millikan and the Twentieth Century Revolution in Physics." *Isis* 68 (1977): 509–526.

Kargon, Robert, and Elizabeth Hodes. "Karl Compton, Isaiah Bowman, and the Politics of Science in the Great Depression." *Isis* 76 (1985): 301–318.

Kármán, Theodore von. *The Wind and Beyond: Theodore von Kármán, Pioneer in Aviation and Pathfinder in Space*. Little, Brown, 1967.

Kevles, Daniel J. *The Physicists: The History of a Scientific Community in Modern America*. Harvard University Press, 1987.

Keyser, C. N. "Aerological Work in the U.S. Navy (Abstract)." *Bulletin of the American Meteorological Society* 1 (1920): 6–7.

Keyser, C. N. "From the Committees." *Bulletin of the American Meteorological Society* 1 (1920): 15–25.

Kimball, Herbert H. "Recent Advances in the Science of Meteorology and in Its Practical Applications." *Bulletin of the American Meteorological Society* 14 (1933): 3–7.

Kimball, Herbert H. "A Review of Recent Advances in Meteorological Research." *Bulletin of the American Meteorological Society* 14 (1933): 185–188.

Koelsch, William A. "From Geo- to Physical Science: Meteorology and the American University, 1919–1945." In *Historical Essays on Meteorology 1919–1995: The Diamond Anniversary History Volume of the American Meteorological Society*, ed. J. Fleming. American Meteorological Society, 1996.

Kohler, Robert. *From Medical Chemistry to Biochemistry: The Making of a Biomedical Discipline*. Cambridge University Press, 1982.

Kohler, Robert. *Partners in Science*. University of Chicago Press, 1991.

Kohler, Robert. *Lords of the Fly: Drosophila Genetics and the Experimental Life*. University of Chicago Press, 1994.

Kushner, David S. "Sir George Darwin and a British School of Geophysics." *Osiris*, second series, 8 (1993): 196–223.

Kutzbach, Gisela. *A Thermal Theory of Cyclones: A History of Meteorological Thought in the Nineteenth Century*. American Meteorological Society, 1979.

Kutzbach, Gisela. "Carl-Gustaf Arvid Rossby." In *Dictionary of Scientific Biography*, volume 11, ed. C. Gillespie. Scribner, 1981.

Kwa, Chunglin. "Modelling Technologies of Control." *Science as Culture* 4 (1994): 363–391.

Lankford, John. *American Astronomy: Community, Careers, and Power, 1859–1940*. University of Chicago Press, 1997.

Larew, Karl. *Meteorology in the U.S. Army Signal Corps, 1870–1960.* Signal Corps Historical Division, 1960.

Lewis, John M. "Cal Tech's Program in Meteorology: 1933–1948." *Bulletin of the American Meteorological Society* 75 (1994): 69–81.

Lewis, John M. "Carl-Gustaf Rossby: A Study in Mentorship." *Bulletin of the American Meteorological Society* 73 (1992): 1425–1437.

Lindzen, Richard S., Edward N. Lorenz, and George W. Platzman, eds. *The Atmosphere— A Challenge: The Science of Jule Gregory Charney.* American Meteorological Society, 1990.

Livingstone, David. *The Geographical Tradition: Episodes in the History of a Contested Enterprise.* Oxford: Blackwell, 1992.

Lowen, Rebecca. *Creating the Cold War University: The Transformation of Stanford.* University of California Press, 1997.

Lynch, P. "Richardson's Forecast Factory: The $64,000 Question." *Meteorological Magazine* 122 (1993): 69–70.

Lynch, P. "Richardson's Marvelous Forecast." In *The Life Cycles of Extratropical Cyclones,* ed. M. Shapiro and S. Grønås. American Meteorological Society, 1999.

Lynch, P. "Weather Forecasting: From Woolly Art to Solid Science." In *Meteorology at the Millennium,* ed. R. Pearce. Academic Press, 2002.

Macrae, Norman. *John von Neumann.* Pantheon Books, 1992.

Malone, Thomas F. "The Atmospheric Sciences and the American Meteorological Society—The More Recent Past." *Bulletin of the American Meteorological Society* 51 (1970): 218–220.

Martin, Geoffrey J. *Ellsworth Huntington: His Life and Thought.* Archon Books, 1973.

Marvin, C. F. "Committee on Research." *Bulletin of the American Meteorological Society* 3 (1922): 11–12.

Marvin, C. F. "The Status, Scope and Problems of Meteorology." *Bulletin of the American Meteorological Society* 4 (1923): 73–76.

McAdie, Alexander. "The Hazard of Sub-Cooled Fog and Ice-Storms in Aviation." *Bulletin of the American Meteorological Society* 10 (1929): 37–38.

McAdie, Alexander. "The Blue Hill Observatory, 1884–1929." In *The Development of Harvard University Since the Inauguration of President Eliot, 1869–1929,* ed. S. Morison. Harvard University Press, 1930.

McCartney, Scott. *ENIAC, The Triumphs and Tragedies of the World's First Computer.* Walker, 1999.

Meisinger, C. LeRoy. "American Doctors of Meteorology and Climatology." *Bulletin of the American Meteorological Society* 4 (1923): 78.

Merton, Robert K. "The Matthew Effect in Science." In *The Sociology of Science: Theoretical and Empirical Investigations*, ed. N. Scorer. University of Chicago Press, 1973.

Miller, Clark A. and Paul N. Edwards. *Changing the Atmosphere: Expert Knowledge and Environmental Governance*. MIT Press, 2001.

Miller, Genevieve. "'Airs, Waters, and Places' in History." *Journal of the History of Medicine* 17 (1962): 129–140.

Millikan, Robert A. *The Autobiography of Robert A. Millikan*. Prentice-Hall, 1950.

Moore, Henry Ludwell. "The Origin of the Eight-Year Generating Cycle." *The Quarterly Journal of Economics* 36 (November 1921): 1–29.

Morrell, J. B. "The Chemist Breeders: The Research Schools of Liebig and Thomas Thomson." *Ambix* 19 (1972): 1–46.

Nebeker, Frederik. *Calculating the Weather: Meteorology in the 20th Century*. Academic Press, 1995.

Nelson, Frederick J. "The History of Aërology in the Navy." *U.S. Naval Institute Proceedings* 60 (1934): 522–528.

Nye, Mary Jo. "National Styles? French and English Chemistry in the Nineteenth and Early Twentieth Centuries." *Osiris*, second series, 8 (1993): 30–49.

O'Brien, Kaye and Gary K. Grice. "Women in the Weather Bureau During World War II." National Weather Service, ca. 2000. http://www.lib.noaa.gov/edocs/women/html.

O'Brien, T. J. "The Navy's Part in Modern Aërological Developments." *U.S. Naval Institute Proceedings* 61 (1935): 385–400.

O'Brien, T. J. "Comments on 'Sixteen Years of American Meteorology and Its Society.'" *Bulletin of the American Meteorological Society* 17 (1936): 380–381.

Oreskes, Naomi. *The Rejection of Continental Drift: Theory and Method in American Earth Science*. Oxford University Press, 1999.

Oreskes, Naomi, and Kenneth Belitz. "Philosophical Issues in Model Assessment." In *Model Validation: Perspectives in Hydrological Science*, ed. M. Anderson and P. Bates. Wiley, 2001.

Oreskes, Naomi, and Ronald E. Doel. "Geophysics and the Earth Sciences." In *The Cambridge History of Science*, volume 5: *Modern Physical and Mathematical Sciences*, ed. M. J. Nye. Cambridge University Press, 2003.

Oreskes, Naomi, Kristin Shrader-Frechette, and Kenneth Belitz. "Verification, Validation, and Confirmation of Numerical Models in the Earth Sciences." *Science* 263 (1994): 641–646.

Owens, Larry. "Science in the United States." In *Science in the Twentieth Century*, ed. J. Krige and D. Pestre. Harwood, 1997.

Palmer, A. H. "Miscellaneous Notes." *Bulletin of the American Meteorological Society* 2 (1921): 71.

Palmer, A. H. "Weather Insurance." *Bulletin of the American Meteorological Society* 3 (1922): 67–70.

Pearce, R. P., ed. *Meteorology at the Millennium*. Academic Press, 2002.

Persson, Anders. "Early Operational Numerical Weather Prediction outside the U.S.A: An Historical Introduction. Part 1: Internationalism and engineering NWP in Sweden, 1952–69." *Meteorological Applications* 12 (2005): 135–159.

Persson, Anders. "Early Operational Numerical Weather Prediction outside the U.S.A: An Historical Introduction: Part II: Twenty Countries around the World." *Meteorological Applications* 12 (2005) 269–289.

Petterssen, Sverre. *Weathering the Storm: Sverre Petterssen, the D-Day Forecast, and the Rise of Modern Meteorology*. Edited by James Rodger Fleming. American Meteorological Society, 2001.

Phillips, Norman A. "The General Circulation of the Atmosphere: A Numerical Experiment." *Quarterly Journal of the Royal Meteorological Society* 82 (1956): 123–164.

Phillips, Norman A. "Carl-Gustaf Rossby: His Times, Personality, and Actions." *Bulletin of the American Meteorological Society* 6 (1998): 1106.

Platzman, G. W. "A Retrospective View of Richardson's Book on Weather Prediction." *Bulletin of the American Meteorological Society* 48 (1967): 514–550.

Platzman, G. W. "The ENIAC Computations of 1950—Gateway to Numerical Weather Prediction." *Bulletin of the American Meteorological Society* 60 (1979): 302–312.

Platzman, G. W. Conversations with Bernhard Haurwitz. National Center for Atmospheric Research, 1985.

Platzman, G. W. "The Rossby Wave." *Quarterly Journal of the Royal Meteorological Society* 94 (1986): 225–248.

Poundstone, William. *Prisoner's Dilemma*. Doubleday, 1992.

Pursell, Carroll W., Jr. "The Anatomy of a Failure: The Science Advisory Board, 1933–1935." *Proceedings of the American Philosophical Society* 109 (1965): 342–351.

Pursell, Carroll W., Jr. "Science Agencies in World War II: The OSRD and Its Challengers." In *The Sciences in the American Context: New Perspectives*, ed. N. Reingold. Smithsonian Institution Press, 1979.

Pyne, Stephen J. *Fire in America: A Cultural History of Wildland and Rural Fire*. Princeton University Press, 1982.

Quiñones, Eric. "Pioneering meteorologist Smagorinsky dies." *News at Princeton*, 3 September 2005.

Rainger, R. "Adaptation and the Importance of Local Culture: Creating a Research School at the Scripps Institution of Oceanography." *Journal of the History of Biology* 36 (2003): 461–500.

Reed, William G. "Meteorological Observations at the University of California." *Science* 37 (23 May 1913): 800–802.

Reed, William G. "Papers Presented at the Toronto Meeting: Military Meteorology." *Bulletin of the American Meteorological Society* 3 (1922): 57–58.

Regis, Ed. *Who Got Einstein's Office: Eccentricity and Genius at the Institute for Advanced Study*. Addison-Wesley, 1987.

Reichelderfer, F. W. "Postgraduate Course in Aerology and Meteorology for Naval Officers." *Bulletin of the American Meteorological Society* 9 (1928): 149.

Reichelderfer, F. W. "Correspondence: The Polar Front Theory and the U.S. Navy." *Bulletin of the American Meteorological Society* 18 (1937): 168–169.

Reichelderfer, F. W. "The Atmospheric Sciences and the American Meteorological Society: The Early Years." *Bulletin of the American Meteorological Society* 51 (1970): 209.

Reiser, Heinz. "The Development of Numerical Weather Prediction in the Deutscher Wetterdienst." In *50th Anniversary of Numerical Weather Prediction Commemorative Symposium, 9–10 March 2000, Book of Lectures*, ed. A. Spekat. Deutsche Meterologische Gesellschaft, 2000.

Richardson, Lewis Fry. *Weather Prediction by Numerical Process*. Cambridge University Press, 1922. Re-issued by Dover, 1965.

Riley, James C. *The Eighteenth-Century Campaign against Disease*. St. Martins, 1987.

Ritchie, David. *The Computer Pioneers: The Making of the Modern Computer*. Simon and Schuster, 1986.

Robie, Bill. *For the Greatest Achievement: A History of the Aero Club of America and the National Aeronautic Association*. Smithsonian Institution Press, 1993.

Roland, Alex. "Science and War." *Osiris*, second series, 1 (1985): 247–272.

Rossby, Carl-Gustav. "Planetary Flow Patterns in the Atmosphere." *Quarterly Journal of the Royal Meteorological Society* 66 (1940): 68–77.

Rossby, Carl-Gustav. "A Message to Members from President Rossby." *Bulletin of the American Meteorological Society* 25 (1944): 268–269.

Rossby, Carl-Gustav. "On the Propagation of Frequencies and Energy in Certain Types of Oceanic and Atmospheric Waves." *Journal of Meteorology* 2 (1945): 187–204.

Rossby, Carl-Gustav, et al. "Relation between Variations in the Intensity of the Zonal Circulation of the Atmosphere and the Displacements of the Semi-Permanent Centers of Action." *Journal of Marine Research* 2 (1939): 38–55.

Sapolsky, Harvey M. *Science and the Navy: The History of the Office of Naval Research.* Princeton University Press, 1990.

Sargent, Frederick, II. *Hippocratic Heritage: A History of Ideas about Weather and Human Health.* Pergamon, 1982.

Schapsmeier, Edward L. and Frederick H. Schapsmeier. *Henry A. Wallace of Iowa: The Agrarian Years, 1910–1940.* Iowa State University Press, 1968.

Scorer, R. S. "Correspondence: Atmospheric Signal Velocity." *Journal of Meteorology* 8 (1951): 68–69.

Scott, James C. *Seeing Like a State.* Yale University Press, 1998.

S. D. F., "Retrenchment in Weather Bureau Activities." *Bulletin of the American Meteorological Society* 15 (1934): 29–30.

S. D. F., "News Item." *Bulletin of the American Meteorological Society* 14 (1933): 212.

Shallett, Sidney. "Weather Forecasting by Calculator Run by Electronics Is Predicted." *New York Times*, 11 January 1946.

Shaw, Sir Napier. "The Outlook of Meteorological Sciences." *Monthly Weather Review* 48 (1920): 34–37.

Shipman, J. M. "Meteorology and Climatology in Normal Schools and Colleges." *Bulletin of the American Meteorological Society* 9 (1928): 86.

Shuman, Frederick G. "History of Numerical Weather Prediction at the National Meteorological Center." *Weather and Forecasting* 4 (1989): 286–296.

Slotten, Richard Hugh. *Patronage, Practice, and the Culture of American Science: Alexander Dallas Bache and the U.S. Coast Survey.* Cambridge University Press, 1994.

Smagorinsky, J. "General circulation experiments with primitive equations, I—The basic experiment." *Monthly Weather Review* 91 (1963): 99–164.

Smith, J. Warren. "From the Committees: Agricultural Meteorology." *Bulletin of the American Meteorological Society* 1 (1920): 19–20.

Smith, Phyllis. *Weather Pioneers: The Signal Corps Station at Pikes Peak.* Swallow Press and Ohio University Press, 1993.

Smith, Richard K. *The Airships* Akron *and* Macon: *Flying Aircraft Carriers of the United States Navy.* U.S. Naval Institute, 1965.

Smith, Richard K. "The Airship, 1904–1976." In *Two Hundred Years of Flight in America: A Bicentennial Survey*, ed. E. Emme. American Astronautical Society, 1977.

Spekat, Arne, ed. *50th Anniversary of Numerical Weather Prediction Commemorative Symposium, Potsdam, 9–10 March 2000, Book of Lectures*. Deutsche Meteorologische Gesellschaft, 2000.

Spence, Clark C. *The Rainmakers: American "Pluviculture" to World War II*. University of Nebraska Press, 1975.

Spengler, Kenneth C. "From the Executive Secretary." *Bulletin of the American Meteorological Society* 27 (1946): 255–256.

Stern, Nancy B. *From ENIAC to UNIVAC: An Appraisal of the Eckert-Mauchly Computers*. Digital Press, 1981.

Stewart, Irvin. *Organizing Scientific Research for War*. Little, Brown, 1948.

Stone, Robert G. "New Department of Meteorology at New York University." *Bulletin of the American Meteorological Society* 19 (1938): 456.

Straley, H.W. III et al. "The Professional Training of Geophysicists: Report of the Geophysical Education Committee, Mineral Industry Education Division." *AIME Transactions* 164 (1945): 397–420.

Taba, Hessam. *The Bulletin Interviews*. World Meteorological Organization, 1988.

Taba, Hessam. *The Bulletin Interviews*. World Meteorological Organization, 1997.

Talman, C. F. "From the Committees: Public Information." *Bulletin of the American Meteorological Society* 1 (1920): 17.

Talman, C. F. "Ice Coating on Aeroplanes." *Bulletin of the American Meteorological Society* 9 (1928): 106–107.

Terrett, Dulany. *United States Army in World War II: The Technical Services: The Signal Corps: The Emergency (To December 1941)*. Department of the Army, 1956.

Thompson, Philip D. "Notes on the Theory of Large-Scale Disturbances in Atmospheric Flow with Applications to Numerical Weather Prediction." *Geophysical Research Papers*, no. 16 (July 1952).

Thompson, Philip D. "A History of Numerical Weather Prediction in the United States." *Bulletin of the American Meteorological Society* 64 (1983): 757–758.

Thompson, Philip D. Interview by William Aspray, 5 December 1986, Charles Babbage Institute, Minneapolis.

Thompson, Philip D. "The Maturing of the Science." *Bulletin of the American Meteorological Society* 68 (1987): 631–637.

Thompson, Philip D. Interview by Joseph Tribbia and Akira Kasahara, 15–16 December 1987, National Center for Atmospheric Research Archives, Boulder.

Townsend, Jeff. *Making Rain in America: A History*. ICASALS Publications, 1975.

Van der Linden, F. Robert. *Airlines and Air Mail: The Post Office and the Birth of the Commercial Aviation Industry*. University Press of Kentucky, 2002.

Van Hise, Charles. Report of the Advisor for Geophysics to the Executive Committee of the CIW, October 1903.

Waldo, Frank. "The Study of Meteorology." *Education* 26 (1906): 149–153.

Walter, Maila L. *Science and Cultural Crisis: An Intellectual Biography of Percy Williams Bridgman (1882–1961)*. Stanford University Press, 1990.

Wang, Jessica. "Liberals, the Progressive Left, and the Political Economy of Postwar American Science: The National Science Foundation." *Historical Studies in the Physical and Biological Sciences* 26 (1995): 139–166.

Ward, Robert D. "How Meteorological Instruction May Be Furthered." *Monthly Weather Review* 46 (1918): 554.

Ward, Robert D. "Are Long-Range Weather Forecasts Possible?" *The Outlook* 140 (1925): 366–371.

Washington, Warren M. "Foreword." In *Historical Essays in Meteorology, 1919–1995: The Diamond Anniversary History Volume of the American Meteorological Society*, ed. J. Fleming. American Meteorological Society, 1996.

Weart, Spencer. *Nuclear Fear: A History of Images*. Harvard University Press, 1988.

Weart, Spencer. "From the Nuclear Frying Pan into the Global Fire." *Bulletin of the Atomic Scientists* 48 (5) (1992): 18–27.

Weart, Spencer. *The Discovery of Global Warming*. Harvard University Press, 2003.

Weber, Gustavus A. *The Weather Bureau: Its History, Activities and Organization*. Appleton, 1922.

Weber, Gustavus A. *The Bureau of Chemistry and Soils: Its History, Activities and Organization*. Johns Hopkins University, 1928.

Weir, Gary E. *An Ocean in Common: American Naval Officers, Scientists, and the Ocean Environment*. Texas A&M University Press, 2001.

White, Graham and John Maze. *Henry A. Wallace: His Search for a New World Order*. University of North Carolina Press, 1995.

White, Richard. *The Organic Machine: The Remaking of the Columbia River*. Hill and Wang, 1995.

Whitnah, Donald A. *A History of the United States Weather Bureau*. University of Illinois Press, 1961.

Winn-Nielsen, Aksel. "Numerical Weather Prediction. The Early Development with Emphasis in Europe." In *50th Anniversary of Numerical Weather Prediction Commemorative Symposium, 9–10 March 2000, Book of Lectures*, ed. A. Spekat. Deutsche Meterologische Gesellschaft, 2000.

Willett, Hurd C. "The Importance of Observations from the Upper Atmosphere in Long-Range Weather Forecasting." *Bulletin of the American Meteorological Society* 18 (1927): 284–287.

Willett, Hurd C. "Carl-Gustaf Arvid Rossby." *Bulletin of the American Meteorological Society* 26 (1945): 243–244.

Williams, Michael R. *A History of Computing Technology*. Prentice-Hall, 1985.

Wobbe, D. "Meteorology in Its Application to Artillery Fire." *Bulletin of the American Meteorological Society* 13 (1932): 137–138.

Woodman, J. Edmund. "The New York University Institute of Aeronautical Meteorology—Its Structure and Problems." *Bulletin of the American Meteorological Society* 17 (1936): 118–119.

Woods, Austin. *Medium-Range Weather Prediction: The European Approach*. Springer, 2005.

Woolard, Edgar W. "Recent Contributions to Mathematical Meteorology." *Bulletin of the American Meteorological Society* 3 (1922): 96–98.

Woolard, Edgar W. "Theories of the Extratropical Cyclone." *Bulletin of the American Meteorological Society* 6 (1925): 49.

Woolard, Edgar W. "The General Problem of Theoretical Meteorology." *Bulletin of the American Meteorological Society* 6 (1925): 78–81.

Woolard, Edgar W. "Conference on Long-Range Forecasting, Cambridge, Mass., Aug. 23, 1937." *Bulletin of the American Meteorological Society* 19 (1938): 123–124.

Woolard, Edgar W. "Charles F. Marvin." *Bulletin of the American Meteorological Society* 26 (1945): 237.

Worster, Donald. *Rivers of Empire: Water, Aridity, and the Growth of the American West*. Pantheon Books, 1985.

Yates, Brig. Gen. D. N. "Remarks Made During the Washington Meeting Discussion on Problems of Industrial and Commercial Applications of Meteorology." *Bulletin of the American Meteorological Society* 28 (1947): 410.

Yeh, Tu-cheng. "On Energy Dispersion in the Atmosphere." *Journal of Meteorology* 6 (1949): 1–16.

Yost, Harold. "Adjusting Rain Insurance Policies." *Bulletin of the American Meteorological Society* 5 (1924): 17–19.

U.S. Committee for the Global Atmospheric Research Program. *The Global Weather Experiment—Perspectives on Its Implementation and Exploitation.* National Academy of Sciences, 1978.

U.S. Weather Bureau. *Report of the Chief of the Weather Bureau, 1919–1920.* Government Printing Office, 1921.

U.S. Weather Bureau. *Report of the Chief of the Weather Bureau, 1920–1921.* Government Printing Office, 1922.

U.S. Weather Bureau. *Report of the Chief of the Weather Bureau, 1921–1922.* Government Printing Office, 1923.

U.S. Weather Bureau. *Report of the Chief of the Weather Bureau, 1922–1923.* Government Printing Office, 1924.

U.S. Weather Bureau. *Report of the Chief of the Weather Bureau, 1923–1924.* Government Printing Office, 1925.

U.S. Weather Bureau. *Report of the Chief of the Weather Bureau, 1924–1925.* Government Printing Office, 1926.

U.S. Weather Bureau. *Report of the Chief of the Weather Bureau, 1925–1926.* Government Printing Office, 1927.

U.S. Weather Bureau. *Report of the Chief of the Weather Bureau, 1930–1931.* Government Printing Office, 1932.

U.S. Weather Bureau. *Report of the Chief of the Weather Bureau, 1932–1933.* Government Printing Office, 1933.

U.S. Weather Bureau. *Report of the Chief of the Weather Bureau, 1933–1934.* Government Printing Office, 1934.

U.S. Weather Bureau. *Report of the Chief of the Weather Bureau, 1934–1935.* Government Printing Office, 1936.

Articles Not Credited to Authors

"The American Meteorological Society." *Bulletin of the American Meteorological Society* 1 (1920): 1.

"Announcements." *Bulletin of the American Meteorological Society* 21 (1940): 306–309.

" 'Bickering' about the Weather." *The Outlook* 142 (27 January 1926): 131.

"Calls Stars Weather 'Key'." *Evening Star* (Washington, DC), 11 October 1934.

"Committees." *Bulletin of the American Meteorological Society* 1 (1920): 5.

"Curiosities of Science and Invention: Meteorology in American Universities." *Scientific American* CV (14 October 1911): 343.

"Effect of Economy Program on U.S. Weather Bureau." *Bulletin of the American Meteorological Society* 14 (1933): 182–183.

"Efficiency of the Weather Bureau Endangered." *Bulletin of the American Meteorological Society* 1 (1920): 140.

"The Electronic Brain." *Weather Bureau Topics* 14 (May 1955): 69.

"Electronic 'Brain' Planned to Forecast the Weather." *Boston Daily Globe,* 12 August 1952.

"Expanding Instruction in Meteorology and Climatology." *Bulletin of the American Meteorological Society* 20 (1939): 206.

"Fog—A Method of Prevention?" *Bulletin of the American Meteorological Society* 11 (1930): 157–158.

"Four New Doctors of Philosophy in Meteorology or Climatology." *Bulletin of the American Meteorological Society* 10 (1929): 166–167.

"From the Committees." *Bulletin of the American Meteorological Society* 1 (1920): 15–25.

"Increased Pay for Weather Bureau Employees." *Bulletin of the American Meteorological Society* 9 (1928): 120.

"Insurance against Adverse Weather." *Bulletin of the American Meteorological Society* 2 (1921): 13.

"Long-Range Weather Forecasting." *Outlook* 140 (20 May 1925): 88.

"Meteorological Data: Progress Report of Special Committee." *Proceedings of the American Society of Civil Engineers* 59 (1933): 153–182.

"Meteorological Data: Progress Report of Special Committee, Discussion." *Proceedings of the American Society of Civil Engineers* 59 (1933): 708–720, 896–900, 1024–1035, 1195–1200, 1340–1343.

"Meteorological Education in the United States: Facilities at Twenty Leading Universities." *Weatherwise* 6 (October 1953): 126–141.

"Meteorology at the Southern Branch of the University of California." *Bulletin of the American Meteorological Society* 2 (1921): 37–38.

"Naval Officers' Advanced Course in Meteorology at the Weather Bureau, Washington, D.C." *Bulletin of the American Meteorological Society* 1 (1920): 90–91.

"Nine and a Half Million Cut from Government Research." *Bulletin of the American Meteorological Society* 13 (1932): 154.

"Notes." *Bulletin of the American Meteorological Society* 14 (1933): 269.

"Notes of Interest to Teachers." *Bulletin of the American Meteorological Society* 1 (1920): 89–91.

"Personal Notes." *Bulletin of the American Meteorological Society* 13 (1932): 238–239.

"Personnel of the Weather Bureau." *Bulletin of the American Meteorological Society* 11 (1930): 71.

"Proceedings of the First Meeting." *Bulletin of the American Meteorological Society* 2 (1921): 13.

"Proceedings of the First Annual Meeting: Report of Committee on Meteorological Instruction." *Bulletin of the American Meteorological Society* 2 (1921): 7–8.

"Science and the Weather's Secrets." *Outlook* 143 (28 July 1926): 428.

"The Second Year of the Society." *Bulletin of the American Meteorological Society* 3 (1922): 1–13.

"Secretary's Report." *Bulletin of the American Meteorological Society* 22 (1940): 34.

"Some Weather Bureau Projects." *Bulletin of the American Meteorological Society* 1 (1920): 11–12.

"The Sun and the Weather." *Literary Digest* 85 (11 April 1925): 25–26.

"Tomorrow's Weather." *Fortune* 47 (May 1953): 144–149.

"University News." *Bulletin of the American Meteorological Society* 3 (1922): 157.

"University Training in Meteorology under Civil Aeronautics Act." *Bulletin of the American Meteorological Society* 19 (1938): 259–260.

"Violence from the Skies." *New York Times* (27 November 1950).

"Weather by Giant 'Brain.'" *Science News Letter* 64 (1953): 309.

The Weather Bureau Is Skeptical." *Outlook* 143 (28 July 1926): 428–429.

"The Weather Bureau Questionnaire on Research Needs." *Bulletin of the American Meteorological Society* 25 (1944): 434.

"Weather Bureau Salaries." *Bulletin of the American Meteorological Society* 14 (1933): 112–113.

Index